The Formation and Future of the Upper Texas Coast

Number Eleven:
Gulf Coast Studies
Sponsored by Texas A&M University–Corpus Christi
John W. Tunnell Jr., *General Editor*

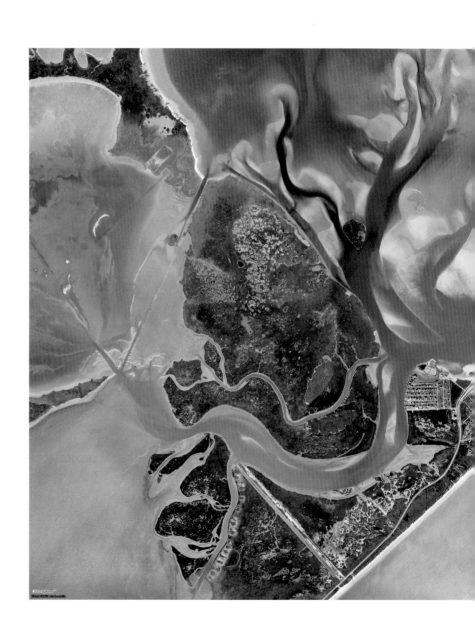

Texas A&M
University Press
College Station

The Formation and Future of the

Upper Texas Coast

A Geologist
Answers Questions
about Sand, Storms, and
Living by the Sea

John B. Anderson

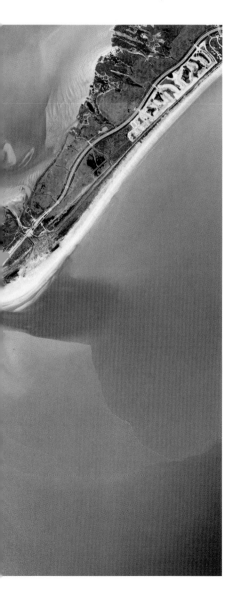

The paper used in this book meets the minimum requirements
of the American National Standard for Permanence
of Paper for Printed Library Materials, Z39.48-1984.
Binding materials have been chosen for durability.
♾

A grant from the Coastal Management Program of the Texas General Land Office helped make
this book possible.

Library of Congress Cataloging-in-Publication Data

Anderson, John B., 1944–

   The formation and future of the upper Texas coast : a geologist answers questions about sand,
storms, and living by the sea / John B. Anderson. — 1st ed.

      p. cm. — (Gulf Coast studies ; no. 11)

   Includes bibliographical references and index.

   ISBN-13: 978-1-58544-561-5 (flexbound : alk. paper)

   ISBN-10: 1-58544-561-4 (flexbound : alk. paper)

   1. Coast changes—Texas.   2. Coastal zone management—Texas.   3. Coast changes—
Mexico, Gulf of.   4. Coastal zone management—Mexico, Gulf of.   I. Title.

GB459.25.A43 2007

551.45′709764—dc22

                              2006024956

Contents

Preface

When I was a child, my family lived for a while on Morgan Peninsula in coastal Alabama. My father, like me, loved the Gulf and could never live far from it. Growing up there I experienced things that piqued my curiosity. For example, our neighbor, who tonged for oysters in the bay, frequently returned with Indian artifacts that he had gathered there. We wondered how these objects occurred in oyster reefs that are now in 5 to 10 feet of water. I now know that these reefs were at one time, just a few thousand years ago, above sea level.

As I grew older, the changes that were taking place along our coast became more evident, and today, a half century later, I occasionally return to our old bay house and view with amazement the amount of shoreline erosion that has occurred during my lifetime. But nature is not the only force that has changed the coast. When I was a kid, miles of beach were accessible to the public. Now I drive along the coast at Gulf Shores, Alabama, for miles and cannot see the Gulf past all the condominiums—at least that was the case until the summer of 2004, when Hurricane Ivan destroyed many of those that lined the coast.

Texans are fortunate. We have vast stretches of undeveloped coastline, and beach access is guaranteed to all citizens by law. Unfortunately these laws are eroding with our beaches.

In 1975 my family and I moved to Houston and immediately began visiting the upper Texas coast. I must admit that at first I was not impressed with Texas beaches. They certainly are not as spectacular as the white sandy beaches of west Florida and Alabama. When we first started going to the beach we saw more cows than people. That is starting to change. The upper Texas coast has finally been discovered, and developers are cashing in on this interest. Preservation of our valued coastal areas hinges on citizens being better informed about how coasts evolve, how they operate, and which natural processes will threaten them in the next century. We must all take an active role in protecting the coast.

I have spent more than two decades studying the Texas coast and continental shelf to better understand how the coast evolved, how the coast will respond to rising sea level and climate change, and how we might best prepare for these changes. Over the years, I have led many field trips for students, teachers, and professionals and given many public presentations on the evolution of the upper Texas coast. My audience's great interest inspired me to write this book. We all feel an attraction to the coast. In the Houston-Galveston area, it is our prime recreational resource. But in our lifetimes we see changes taking place that concern us, and we want to understand why these changes are occurring. How much is due to human influence, and how much is simply natural evolution?

I wrote this book because I feel that time is running out to save our coast. I see too many things happening that will have a long-term detrimental impact. My goal is to impress upon readers the fact that our coast is a fragile system that requires our immediate attention. One of the challenges of writing a book like this is that if it is too heavy on the science, people will not want to read it. If the science is omitted, some will say that the book is heavy on opinion and light on supporting data. I have tried to get around this problem by providing important background information needed to understand scientific issues. I have included some key figures that present hard data, but these are not critical to understanding the basic processes that are occurring along our coast. I have also included a fairly comprehensive list of references for those who would like to learn more or locate supporting data for my interpretations.

Acknowledgments

I want to thank my graduate students who conducted research on the Texas coast and continental shelf. This book is based largely on their hard work. Thank you, Ken Abdulah, Laura Banfield, Lou Bartek, Mary Lou Cole, Brenda Eckles, Michelle Fassell, Mike Hamilton, Jessie Maddox, Kristy Milliken, Tony Rodriguez, Alex Simms, Fernando Siringan, Wendy Smyth, Jennifer Snow, Patrick Taha, Mark Thomas, and Julia Wellner. A list of these students' theses and dissertations is provided in the bibliography. In particular, I am indebted to Tony Rodriguez, who has been my partner in Gulf Coast research for the past decade. April Metz improved many of the figures and photographs for this book. Our research was funded by a consortium of oil companies, including BP/Amoco, Agip, Anadarko, Chevron, Conoco/Phillips, Exxon/Mobile, Marathon, and Shell. The National Science Foundation (NSF) also supported our work, and we are especially grateful to Dr. Rich Lane of NSF for recognizing the need for this type of research and for encouraging me to make the results of our work available to the general public. Our study of sand resources was funded by the Texas General Land Office. Thanks to Dr. Juan Moya of the General Land Office (GLO) for his support and encouragement and to his colleagues Tom Calnan, Ray Newby, Kathy Smartt, and Sheri Land for reviewing the manuscript. Several of the graduate students who worked on Gulf Coast projects were funded by the David Worthington Graduate Fellowship. Thank you David and Beverly for your support and encouragement. My wife, Doris, was my field partner on many excursions and provided literary advice. She also gave up many weekends while I wrote.

Lastly, I wish to thank my son John, who spent many long, hot days in the field with me collecting data. This book is dedicated to him.

Introduction

Ask any coastal geologist whether natural forces, particularly sea level rise, or humans are having the greatest impact on the world's coasts, and they will probably say that humans are winning the race. Factor in our contributions to global warming and coastal subsidence, or sinking of the land surface, and we, by far, are the greatest threat to coasts. Nowhere on Earth does the impact of humans and natural forces pose a greater threat to coasts than in Louisiana and Texas.

This book deals with the part of the Texas coast between Sabine Pass and the Brazos River. This part of our coast is the most populated and the most threatened by coastal subsidence and overdevelopment. As a result, it has been more greatly affected than any other part of the Texas coast.

Those of us who have visited the beach for more than a decade, a time interval that now seems short to me, need little convincing that our shoreline is in a state of change. A seawall stands as a fortress between the city of Galveston and the advancing shoreline. Long-term shoreline retreat along the upper Texas coast has occurred at rates between 3 and 15 feet per year. The beach that exists seaward of the seawall is artificial, nourished by sands pumped from offshore during the mid-1990s. County Road 87 between Galveston and Sabine Pass has been overtaken by the advancing shoreline, and all along the coast the advancing shore is consuming houses. Phase one of the Galveston Seawall, constructed between 1903 and 1922 after the "Great Storm" of 1900, has been called the mightiest seawall on any barrier island in the world. It now stands as the last line of defense against the advancing shoreline.

Other changes taking place within our coastal system are not as obvious but pose an equal threat to our coasts. The shorelines of area bays are eroding at alarming rates, averaging just over 2 feet per year, but in places exceeding 10 feet per year (Bureau of Economic Geology Web site, http://www.beg.utexas.edu/coastal/coasta101.htm; see also chapter 5).

Phase one of the Galveston Seawall was constructed between 1903 and 1922, after the Great Storm of 1900, and has been called the mightiest seawall on any barrier island in the world.

West of the Galveston Seawall the shoreline continues to advance landward as sea level rises. The remains of the foundation for a VHF antenna that once stood behind the dune line is a reminder of this ongoing process.

The eastern shoreline of Galveston Bay has become a virtually continuous line of concrete riprap that helps slow the erosion of the shore. The wetlands that once lined the eastern shore are gone.

This photograph of County Road 87 was taken in 1977. The highway, which once connected High Island and Sabine Pass, was overtaken by the advancing shoreline and was closed a few years after this photo was taken. Today County Road 3005 on Follets Island faces a similar fate.

In addition, thousands of acres of wetlands have been submerged by subsidence in the past fifty years. Humans have contributed to coastal and bay erosion and wetlands loss by damming rivers, increasing rates of land subsidence with groundwater withdrawal and oil and gas extraction, and developing land without allowing for the natural migration of coastal environments. The construction of ship channels and the Texas City Dike has altered the natural circulation within bays. This in turn has altered the salinity of the bays and affected their inhabitants. The laws that are intended to protect the coast are, for various reasons, not working.

The changes that have occurred along the upper Texas coast may well be minimal compared to those changes that will occur this century as our climate continues to change and the rate of relative sea level rise increases. Chapter 1 of this book explains some of the basic geological processes that occur daily and control those changes that we currently observe along our coast. Chapter 2 focuses on the long-term evolution of our coast, a perspective that allows us to predict future changes and to gauge human impact relative to natural changes. Chapter 3 reviews historical changes along the upper Texas coast and human impact on the coast.

West of San Luis Pass, County Road 3005 is slowly being overtaken by the advancing shore. This stretch of highway probably would not survive a hurricane. When the highway is destroyed, there is no place to rebuild it without infringing on wetlands.

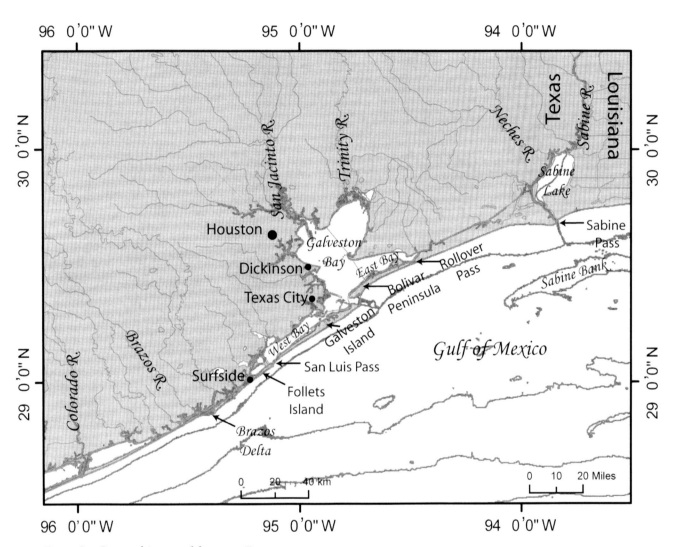

Figure I.1. Geographic map of the upper Texas coast.

Chapter 4 examines reasons that sea level is rising and theories that predict the rate of rise will increase, and perhaps is already increasing. This chapter also explores the issue of coastal subsidence and human influence on this process.

Most methods used to combat coastal erosion have not worked and have even been counterproductive, the subject of chapter 5. The only way to combat beach erosion, short of extending the Galveston Seawall along the entire stretch of our coast, is to nourish beaches with sand. But sand is in short supply on the upper Texas coast. During 2004 and 2005 an unprecedented number of powerful storms impacted the Gulf Coast. Was this an unusual phenomenon, or are we entering a phase of increased storm frequency and magnitude related to global warming? How do these events compare to the long-term record of storm impact, and what would be the impact of a Category 4 or 5 hurricane on the more populated areas of the coast? These questions are addressed in chapter 6.

Chapter 7 discusses potential sources of sand for beach nourishment and some of the challenges involved with nourishing our beaches. Finally, Chapter 8 addresses some concerns related to continued coastal development. This chapter also outlines existing laws for coastal preservation and lists those government agencies that are responsible for safeguarding the coast.

This book was written as a wake-up call for everyone who cares about coasts. If we do not act now, our grandchildren will see changes to our coasts that many of us cannot even comprehend. Figure I.1 is a geographic map of the area that will help the reader locate places mentioned in the book.

The Formation and Future of the Upper Texas Coast

Coastal Processes

View of the beach during low tide.

What happens to the sand that erodes from our beaches?

Wave Motion and Sand Movement

As waves approach the shore, they begin to drag on the seafloor, which causes them to become steeper and break. A breaking wave is literally surface water that has overrun water near the bottom. The location where waves break is called the breaker zone. Its location varies as wave height varies. Landward of the breaker zone is the surf zone, where extreme turbulence from breaking waves lifts sand off the seabed. The area where waves swash back and forth onto the beach is called the swash zone (fig. 1.1).

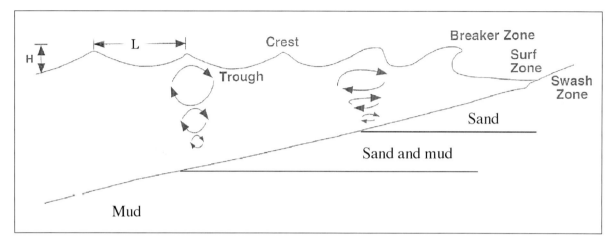

Figure 1.1. When waves approach the beach they become steeper as they drag on the bottom. Eventually the surface water overruns the bottom water, and the wave breaks. As the height of the wave (H) and the length of the wave (L) increases, the wave breaks farther from shore.

Landward of the swash zone is the main part of the beach. Coastal geologists refer to this as the storm beach because it becomes a swash zone during storms and extremely high tides. Sand that is moved landward by swash and wind piles up at the landward side of the beach to create a beach ridge, which includes dunes. This is the natural barrier that protects the coast from storm wash-over. As the beach retreats landward, the dune line moves with it. Unfortunately, construction on the beach has too often prevented this natural movement of the dune line, so it is becoming a rarity along our coast.

The motion of water within a wave is orbital, or circular, and the upward component of wave motion lifts sand grains off the seabed each time a wave passes. This is why waves are so efficient in transporting sand. The depth of wave orbital motion increases with wave height and length (the distance between wave crests), meaning that larger, longer waves, in essence, dig deeper, transporting sand from greater depths than do smaller waves. As waves approach the beach they lift sand grains off the bottom and toward shore. If you stand in the breaker zone you can feel this motion of sand from the breaking wave. The sand that is transported shoreward is deposited as a sandbar. As the bar grows, waves break over it, so the breaker zone becomes more confined. The location of the breaker zone depends upon the wave length and height on any given day. When onshore winds (those that move from water toward shore) are strong, large waves

Storm beach. During an extreme high tide, the wave swash extends to the base of the dunes.

View looking west on Galveston Island showing the dune line. Features like this are becoming more and more rare because humans have interfered with their natural landward migration.

Two sandbars, one near the beach and an outer bar where the waves are breaking.

break farther offshore to form sandbars farther offshore. Surf fishermen are well aware that there is more than one sandbar off the beach, each separated by depressions that run parallel to the coast. The depressions, or runnels, are where the fish tend to congregate.

Sandbars form far from shore during strong winds, when waves are larger. These offshore bars tend to endure, because when winds diminish, the waves are too small to have much influence on the bars. On any given day there may be as many as three sandbars and associated breaker zones. Over periods of weeks, as the weather and wave conditions vary, sandbars migrate landward and old bars are remolded into new ones. If you were to measure the seafloor depth from the beach to a few hundred yards offshore on a regular basis, you would observe constant change in

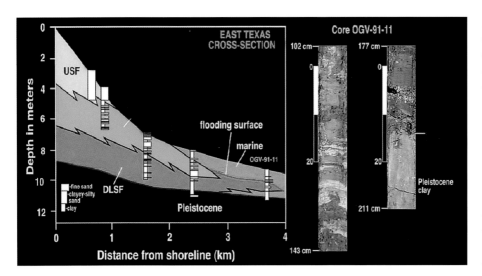

Figure 1.2. Sediment cores collected offshore of Galveston Island State Park show the change from sand to mud in an offshore direction. They also show that mud (in orange) now buries sand as a result of coastal retreat. The photographs on the right are examples of sediment cores used to construct this profile. The lighter units are sand, which is very fine grained and unsuitable for beach nourishment. The darker layers are mud, which further limits the quality of the sediment for beach nourishment. Modified from Rodriguez, Fassell, and Anderson 2001.

the location of runnels and bars. This highly dynamic near-shore zone is referred to as the upper shoreface. This part of the coast is under the constant influence of waves, and only sand is deposited here. It is also the steepest portion of the shoreface and, along the upper Texas coast, occurs between the shoreline and about 15 to 18 feet (5 to 6 meters) water depth.

Seaward of the wave-dominated upper shoreface is a zone that is influenced by waves only during storms. During storms, sand is delivered to this zone, the lower shoreface, from onshore. During fair weather conditions, the seafloor is more quiescent and mud is deposited. With time, sand mixes with mud. Along the upper Texas coast, the lower shoreface extends seaward of the upper shoreface to water depths of about 24 to 30 feet (8 to 10 meters) (fig. 1.2). The depths of these zones and their sediment types differ largely according to differences in the bottom profile along the coast. Seaward of the lower shoreface is the continental shelf.

For those of us concerned about coastal erosion and beach nourishment, it is important to know where within the coastal setting sand is being transported by waves and coastal currents. The term *closure depth* is commonly used to describe the maximum depth of sand transport. One way to estimate closure depth is to use such factors as wave length and height. However, in my opinion, the best way to determine the closure depth is to use the actual distribution of sand in the shoreface and to observe the migration of sandbars over time.

The coastal waters of the upper Texas coast tend to be muddy much of the time because larger waves resuspend mud that rests on the seafloor just offshore. This sediment is then moved alongshore by coastal currents.

Longshore Currents

The constant movement of waves toward the beach causes water to pile up at the coast, and that water ultimately has to move back offshore. But before it does, the water may flow parallel to the shore as a longshore current. This happens when waves approach the beach from an angle. Longshore currents can be quite strong if the winds are strong and the waves approach from a sharp angle. These longshore currents accelerate as more water is continuously added to the coastal current.

We all have experienced longshore currents. Recall the time you were drifting on your float and suddenly realized that you had drifted down the coast some distance from where you entered the water. You were taken there by longshore currents. The experience could have been much worse, because ultimately longshore currents veer offshore as waves move more water onshore. This offshore flow occurs either as discrete currents known as rip currents or as a more dispersed flow commonly referred to as undertow. These offshore-directed currents pose a danger to swim-

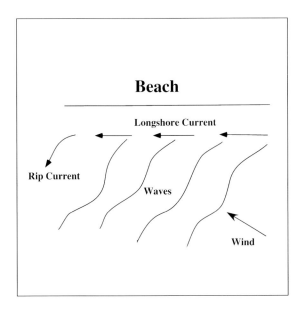

Figure 1.3. Waves that approach the coast from an angle force water onshore and alongshore to produce longshore currents.

mers, who often panic when they feel themselves being pulled offshore. The trick is to remain calm, because the current will dissipate when it flows a short distance offshore and its water is displaced.

Along the upper Texas coast, prevailing winds are from the southeast. This creates waves that approach the coast from the southeast and longshore currents that flow from east to west. These longshore currents transport sand, and the constant movement of sand along the coast is called longshore transport. The quantity of sand moved within the longshore transport system is evident wherever longshore currents are blocked by man-made structures. For example, at the west end of Bolivar Peninsula, sand transported to the west by prevailing longshore currents is trapped by the North Jetty, which extends 4.5 miles offshore. It is estimated that the jetty has trapped 28 million cubic yards of sand since it was constructed. Another way to visualize the volume of sand moving within the longshore transport system is to imagine dump trucks traveling west along Galveston Island filled with sand. A new load of sand would pass every twenty minutes.

From November through March, low-pressure weather systems (fronts) frequently move across the upper Texas coast. When they do, we experience strong offshore winds (those that blow from the shore out to the sea). These winds dampen waves, and longshore currents are weak.

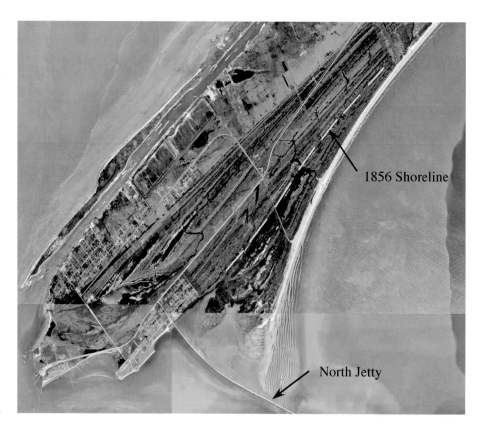

Figure 1.4. Along Bolivar Peninsula, the longshore drift system delivers sand toward Bolivar Roads, the pass between Galveston Island and the peninsula. But before the sand reaches the inlet it is blocked by the North Jetty. If you drive to the area just east of the jetty, you can see the extensive beach that has formed from sand trapped on the "up-drift" side of the jetty. Note the location of the 1856 shoreline (dashed line) before the jetty was constructed. Note also the linear features that extend along the peninsula. These beach ridges, separated by depressions or swells, formed when the peninsula was growing naturally. Photo from the U.S. Geological Survey.

Shifting northerly and southeasterly winds enhance tides, and beaches expand and contract accordingly.

On Galveston Island, as fronts move across the coast and to the east, winds blow from the west, and longshore currents flow toward the east. Although this is not the prevailing longshore current direction, sand is trapped behind the jetties on the east end of the island, where it remains because the jetties prevent waves and longshore currents from removing this sand. This is why East Beach is so expansive.

The Bolivar jetties were constructed to protect ships entering the ship channel and to prevent sediment from filling the channel. Before the jetties were constructed, sand was transported through the Bolivar Roads inlet into Galveston Bay or offshore, where it accumulated as an extensive bar off the mouth of the inlet. The sand delivered into the bay accumulated in a sand body that is referred to as a flood tidal delta, because the

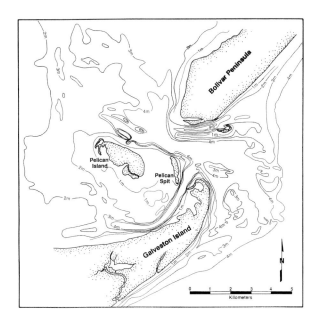

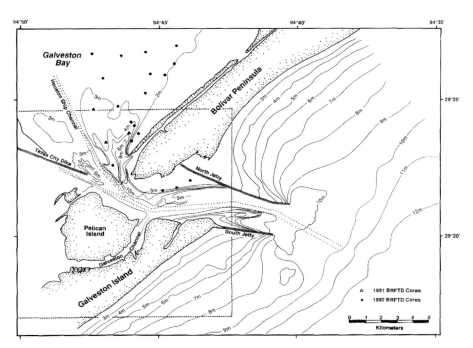

Figure 1.5. After the turn of the twentieth century, jetties were constructed to provide protection for vessels entering the ship channel at Bolivar Roads. The chart at the top is from 1856, and the one at the bottom is current. Note the changes that have occurred in the ebb and flood tidal deltas since the jetties were constructed.

Rock groins that extend offshore of the Galveston Seawall capture sand that moves within the longshore transport system.

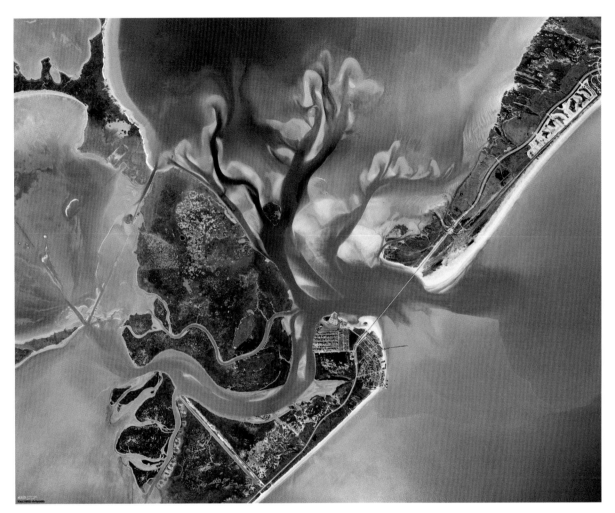

Aerial view of the San Luis Pass tidal delta. Photo from GlobeXplorer.

sand is delivered there during the rising or flood tide. Pelican Island was originally part of the flood tidal delta. The offshore sand deposit is referred to as the ebb tidal delta because the offshore-directed or ebb tide delivers the sand to these areas. After the jetties were constructed, the ebb tidal delta was eroded, which is evident in the differences in offshore bathymetry (water depth) between the two charts in figure 1.5. The flood tidal delta was mostly buried in mud.

Along Galveston Island, sand is mostly transported to the west. But the longshore drift has been altered by rock groins that extend offshore of the

The beaches east of San Luis Pass have experienced significant growth and retreat in recent years. In the recent past they have eroded rapidly. This area is part of the tidal delta and, as such, is subject to constant change.

seawall. The groins were built to trap sand moving within the longshore drift system and to slow the rate of beach erosion. They have also slowed the rate of sand delivery to beaches west of the Galveston Seawall. Actually, so little sand is currently moving in the longshore transport system that removing the rock groins would have little effect on coastal erosion.

Most of the sand eroded from Galveston beaches is transported westward and ultimately ends up in the San Luis Pass tidal delta. You can see these sand accumulations from the bridge at San Luis Pass. Or, if you are a boater, you may have run aground on one of these bars. With time, the tidal inlet and delta migrate toward the west, the direction of longshore transport. Sediment cores collected to the east of the modern tidal delta within West Bay have sampled sand from the former delta that is now buried beneath bay mud.

San Luis Pass is one of the few remaining natural tidal inlets on the Texas coast. As such, it has a history of westward migration that occurs as more and more sand is delivered there by the westward longshore transport system. The beaches adjacent to the ebb tidal delta have a history of constant fluctuation as the tidal delta alternately shifts landward and seaward. As the deep tidal inlet migrates toward the west, it undermines houses.

The beaches around San Luis Pass are among the most rapidly changing Texas beaches because sand is constantly shifting from the beach into the offshore bars and vice versa. Here, houses at Treasure Island are literally sitting seaward of the surf zone, having been left behind by the ever-shifting ebb tidal delta.

West of San Luis Pass, along Follets Island and around Surfside Beach, the coast has a low profile; Surfside has an average elevation of 4 feet. The most significant loss of sand from the beach occurs during storms when sand is washed across the beach into the wetlands and back-barrier bays. Eventually, this sand will be reclaimed and moved back into the long-shore transport system as the shoreline advances landward. Meanwhile, the sand helps to maintain wetlands by providing a framework on which marsh vegetation grows.

Aerial view west of the Brazos Delta after Tropical Storm Frances. Note how sand has been washed across the beach into the lagoons and wetlands.

What happens to the area immediately offshore of the beach as the shoreline moves landward?

The fact is, the beach is not the only part of the coast that is retreating landward. The shoreface is also retreating, and at the same rate as the beach. This is because nature works to maintain a constant beach and shoreface profile as the coast retreats landward. Coastal geologists refer to this as the equilibrium profile.

Figure 1.6 shows two of the several geological models for how the shoreface and beach move landward. Note that a constant shoreface profile is maintained. The only exception occurs when the rate of sea level rise changes or when the amount of material being eroded by waves changes

In its natural state, the beach will maintain a constant profile as it retreats landward. However, if the beach is not allowed to migrate landward, it grows narrower and its profile grows steeper. That is the case here, where a bulkhead was constructed to stop coastal retreat. The result is a more unstable beach, and the houses on the beach are highly susceptible to storm undercutting. The stairs in this photograph have been undercut by erosion, leaving them suspended one foot above the beach.

This photograph illustrates the degree to which the beach profile is lowered with time. In this case, the metal bulkhead is approximately 4 feet high, so the beach profile has been lowered that much. With time, or during a major storm, the bulkhead will be destroyed, and the natural beach profile will be reestablished, undercutting the houses behind the bulkhead.

Figure 1.6 a and b. These two models attempt to capture how the shoreface and beach migrate landward in response to relative sea level rise. Note that in model A, a constant shoreface profile is maintained during retreat. Sand that is eroded from the shoreface is either moved landward or remains in the longshore transport system. Model B illustrates the manner by which the shoreface moves landward and cuts a new profile. The shape of this profile is controlled by the rate of sea level rise and by sediment supply. During a fast rise, or a significant reduction in sediment supply, steps are created in the profile as the shoreface shifts rapidly landward. Changes in the offshore profile also occur where waves encounter different substrates that are either easier or harder to erode. Modified from Brunn 1962 and Swift 1975, respectively.

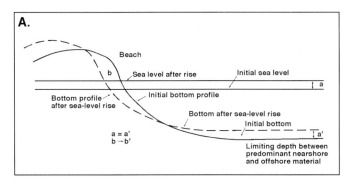

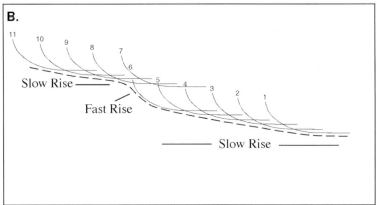

significantly. Sand that is eroded from the beach and upper shoreface moves landward or remains in the longshore transport system. This is why it makes little sense to dredge sand from the shoreface to nourish a beach. Doing so increases the slope of the shoreface, creating an unstable profile that must be compensated for by the offshore movement of sediment until the stable profile is reestablished.

As the shoreface migrates landward, an erosion surface is left behind on the shelf. Compare the models in figure 1.6 with the actual profile shown in figure 1.2. Note that at a distance of 4 kilometers from the beach, the sandy sediments that composed the beach and inner shoreface have been entirely eroded, almost as though a huge bulldozer had plowed its way landward, removing anything above the base of the shoreface. That is exactly what storm waves do to the shoreface and beach. This process is called shoreface ravinement, and it produces a surface where marine mud rests directly on deposits that were laid down many thousands of

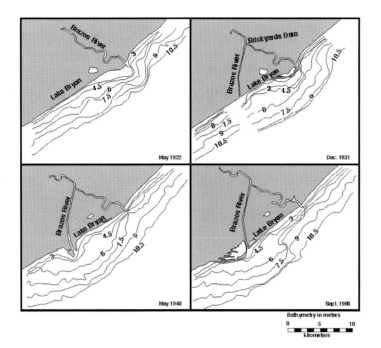

Figure 1.7. These charts show the Brazos Delta before and after the Brazos River was diverted in 1929. Note that a prominent delta (shallow area) existed offshore of the river before it was diverted. Note also the westward shift of the delta after the river diversion, which indicates that the sand that was eroded from the old delta was redeposited in the new delta.

years ago in the Pleistocene. This surface is called the shoreface ravinement surface. Off the west Florida and Alabama coasts, sand is so plentiful that the shoreface profile does not erode below the level of these sands. That is why Floridians have been able to nourish their beaches with sand from the area directly offshore of the shoreface.

Based on the models shown in figure 1.6, any part of the coastline that is situated above the shoreface ravinement surface is destined to complete destruction as the shoreline advances landward. Sediment cores from offshore confirm these models (fig. 1.2). Most of Galveston Island and Bolivar Peninsula are resting above this surface and will be destroyed in coming centuries. This is also why there is little in the way of an offshore record of old barrier islands that existed on the shelf prior to 5,500 years ago. They were removed by shoreface ravinement. Sand banks are an exception; those are discussed in chapter 2.

One of the most impressive characteristics of the north Texas coast is the rapid rate at which shoreface ravinement occurs. For example, figure 1.7 shows old bathymetric maps for the area offshore of Surfside Beach, where the Brazos Delta was located before the river diversion of 1929. Note that the delta was a prominent lobe, marked by shallow depths, offshore of the river mouth. Within a few decades after the river was diverted, the lobe was gone, having been eroded by storm waves, and a new delta lobe was formed to the southwest, offshore of the new Brazos River mouth. This is one of several examples of the fact that sediment that occurs offshore in water depths of up to 30 feet is eventually exhumed as the shoreline advances. Much of the sand that is exhumed is delivered back into the longshore transport system. In this case, the delta was removed in about two decades. Another example is the erosion of the old Bolivar Roads ebb tidal delta that occurred after the construction of the ship channel and jetties, which blocked sand supply to the tidal delta (fig. 1.5). The lesson here is that it doesn't take long to see the impact of human tampering with the coastal system.

Now let's get back to the question of where the sand eroded from area beaches goes. Most of the sand that erodes from Bolivar Peninsula is trapped on the beach east of the North Jetty. The sand eroded from Galveston Island ends up in the San Luis Pass tidal delta. Sand eroded from Follets Island washes over this narrow barrier into wetlands and bays. A similar fate awaits sands removed from beaches between the Freeport jetties and the Brazos Delta. During major storms, sand can be removed from the beach and transported far offshore and deposited as storm beds. However, sediment cores taken offshore have rarely sampled storm beds, so this mechanism of sand removal from the coast does not appear to be significant.

As for what happens to the area immediately offshore of the beach as the shoreline moves landward, the answer is pretty well known. As the shoreline retreats landward, the shoreface retreats with it, leaving little remaining shoreface deposits on the inner continental shelf. Sand is confined to the area closest to shore, generally within a kilometer (0.6 miles) of shore and within the steepest, most dynamic portion of the shoreface. The rate of shoreface movement is more or less equal to changes that occur on the beach. If sand is taken from the shoreface for beach nourishment, an unstable offshore profile is created and sand will be delivered from the nourished beach back to the shoreface to reestablish the equilibrium profile.

The Evolution of the Modern Coastal System

Aerial view of the upper Texas coast. Photo from GlobeXplorer.

How did our coast evolve?

To appreciate why our coasts are changing we need to understand how they evolved. To do that, we need to go back in time thousands of years to when the Texas shoreline was located many tens of miles seaward of its present location.

Between 120,000 and 20,000 years ago the Earth's climate shifted into a cold phase, and ice sheets began to expand in Europe, North America, Greenland, and Antarctica. In the Northern Hemisphere the Pleistocene ice sheets extended far down into Europe, and in North America the ice sheet covered virtually all of Canada and much of the northern United States. During that time interval the Brazos, Colorado, and western Louisiana rivers constructed vast deltas, as large as the present Mississippi River Delta, that advanced across the continental shelf as sea level fell. To the south, the Rio Grande also formed a large delta. Offshore of these deltas the water depth dropped to more than 200 feet within 10 miles of shore. The only place where extensive sandy beaches existed was on the central Texas coast, where there are no large rivers. During this time rivers supplied much of the sand that now comprises our coastal system.

By 22,000 years ago, the ice sheets had reached their maximum proportions. The water that comprised these great ice sheets came from the oceans, so much water that global sea level was approximately 350 feet (120 meters) below its present level. The east Texas shoreline was situated nearly 80 miles south of its current position, and rivers incised and extended their valleys to the edge of the continental shelf. Approximately 18,000 years ago, the great ice sheets began to melt as the Earth's climate warmed. The water from these melting ice sheets raised the sea level to its present position, and the shoreline began its landward migration.

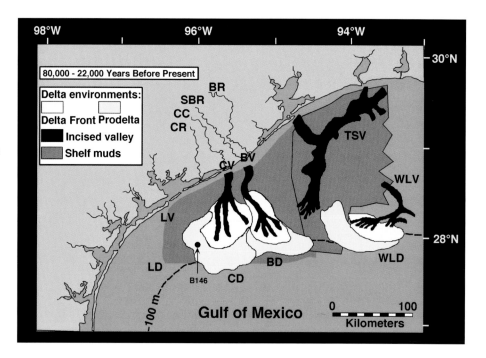

Figure 2.1. The locations of the Colorado, Brazos, and western Louisiana deltas, which existed offshore of the upper Texas coast between about 80,000 and 22,000 years ago. Modified from Anderson et al. 1996.

Geologists who have studied the rise in sea level from glacial melting have constructed curves that show the rate of sea level rise over time. The most widely accepted curves were derived by radiocarbon and uranium/thorium age dating of fossil corals, which are known to live at or near sea level, in drill cores from places like Barbados, Tahiti, and New Guinea. These islands were chosen because they are not subsiding and thus are good benchmarks for sea level change. The sea level curves from these areas provide an important framework for understanding how coasts evolved, and geologists use them to help reconstruct episodes of coastal evolution.

Rising Sea Level and Drowning Coasts

In east Texas the shoreline has retreated landward approximately 80 miles in the past 17,000 years. As the shoreline receded, it cannibalized the old deltas that had formed as sea level was falling. The sand from

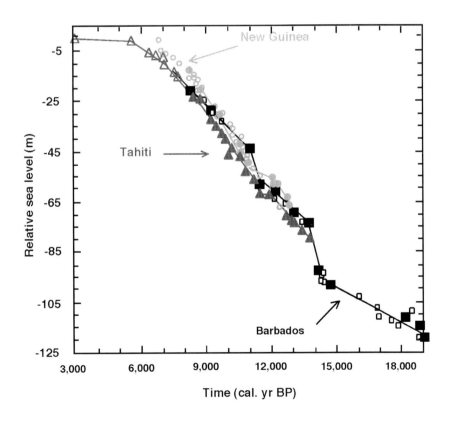

Figure 2.2. The sea level record used by geologists was obtained from drill cores collected from coral islands such as Barbados, Tahiti, and New Guinea. Radiocarbon and uranium/thorium ages from corals that are known to have lived within about 15 feet of sea level show the rise in sea level over the past 18,000 years. Modified from Bard et al. 1996.

these deltas became part of the newly evolving coast. The rate of sea level rise was greatest between 14,000 and 5,000 years ago, when ice sheets retreated most rapidly. During this interval, the shoreline moved landward in some locations as much as 60 feet per year, but this rate was far from constant. At times the coast moved landward several miles within a few centuries. A number of large sand banks offshore, including Sabine Bank and Heald Bank, attest to these rapid changes. Sediment cores from these banks contain shells of organisms that today live at or near the shoreline, one piece of evidence that these banks are old barrier islands that were literally drowned by the rising sea. Radiocarbon dates from shells collected within the offshore banks provide insight into when these former barriers existed. Heald Bank, for instance, was a barrier island approximately 8,000 years ago. Sabine Bank, on the other hand, is much younger. Shells from Sabine Bank range in age from just over 5,000 years old to as young as 2,500 years old. Today these banks rest in water depths of 25 to 50 feet. They are the grave markers of former barrier islands.

Figure 2.3. Research in recent years has led to a detailed reconstruction of coastal evolution for the offshore area between Sabine Pass and San Luis Pass. This figure shows the location of the shoreline at different times in the past (ka=1,000 years). It also shows the old Sabine and Trinity river valleys (in gray) that were flooded to create Sabine Lake and Galveston Bay, respectively. These former shorelines indicate rates of coastal retreat of 20 meters (60 feet) per year to 5 meters (15 feet) per year during this time interval between 7,700 BP (7.7 ka) and 2,800 BP (2.8 ka), compared to the current rate of 1 to 1.5 meters (3 to 5 feet) per year. Offshore banks are shown in light grey. The outer bank is Heald Bank and the inner bank is Sabine Bank. From Rodriguez et al. 2004.

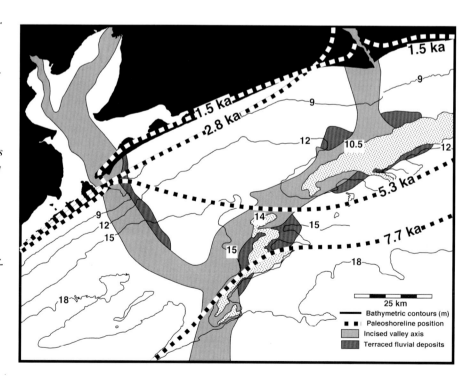

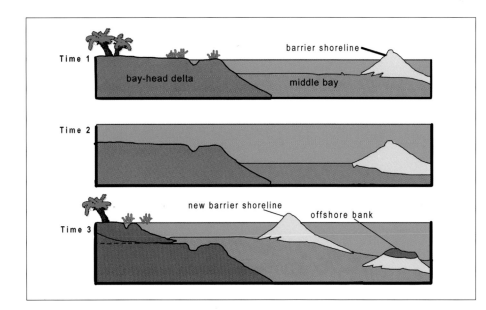

Figure 2.4. This figure illustrates how old barrier islands were drowned in place by rising sea level to form offshore banks such as Heald Bank and Sabine Bank. Courtesy of A. Rodriguez.

Galveston Island's Evolution

How and when did Galveston Island form?
Has it always retreated landward?

Barrier islands, which formed in the past few thousand years as the rate of sea level rise decreased, are common features along the Gulf Coast. Coastal geologists have debated their origins, but the evolution of Galveston Island, one of the best studied barrier islands in the world, is well established. The island began to form about 5,500 years ago, and for most of its history it grew and advanced seaward. The island's growth left an imprint on the landscape that can be seen in aerial photographs, similar to coastal landscapes along many of the world's shorelines. This landscape consists of linear ridges and swells between these ridges. The ridges are the highest natural features on the island, and the swells provide most of the island's limited freshwater habitats.

The sand ridges of Galveston Island contain shells that include the small coquina clam, *Donax*. *Donax* lives today along the Texas shore but only within the swash zone. Hence, those ridges that contain their shells are old beaches. Radiocarbon ages from *Donax* shells within ridges show

Aerial view of Galveston Island. Photo from GlobeXplorer.

Figure 2.5. This aerial view of the central part of Galveston Island shows linear, east-west oriented ridges and swells that characterize the island's topography. The channels that cut through the back side of the island were formed during hurricanes. Note that they do not cut across the beach ridges on the Gulf side of the island, which tells us that the island has not been breached by hurricanes since the younger, more seaward ridges were formed. Photo from the U.S. Geological Survey.

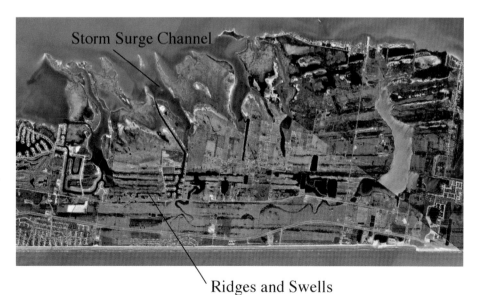

Storm Surge Channel

Ridges and Swells

Ground-level view of ridge and swell on Galveston Island's west end.

Water-filled swells provide freshwater habitats and drinking water for the island's wildlife.

Donax is a small clam that lives in the swash zone, the area where waves swash back and forth onto the beach. Since these clams live at the shoreline, when they are found in older deposits, we know that these deposits were formed at sea level.

that the ridges decrease in age in an offshore direction. So the ridges reflect the island's steady offshore growth.

Using beach ridges and radiocarbon dates from them, geologists have been able to reconstruct with detail the evolution of Galveston Island. In the 1950s, Dr. Rufus Le Blanc and his colleagues from Shell Oil Company first studied the island's formation. They felt that ancient barrier islands that are now buried deep in the subsurface might contain significant oil and gas reserves, and they wanted to learn more about how these sand bodies are formed. Their study was limited by the rather primitive state of radiocarbon age dating at the time, although it did show continuous growth of the island as the rate of sea level rise began to decrease. The island grew seaward by adding new layers of sand that was being transported from offshore as the shoreline steadily advanced landward; as the island grew, rising sea level drowned the low-lying area behind the island to create West Bay.

About 2,600 years ago, the island was only half as wide as it is today. Until then it was frequently breached by storms, which created the storm surge channels that occur on the back side of the island (fig. 2.5). Beach ridges that are younger than about 2,600 BP have not been breached. Thus the east end of the island became wide enough to avoid breaching after that time. The narrower west end of the island is still susceptible to breaching during a major hurricane (see chapter 6).

The beach ridges that record the most recent phase of Galveston Island's growth are now being overtaken by the advancing shoreline. Note that the youngest beach ridges on the east end of the island are about 1,800 years old. Younger ridges have been eroded by the advancing shoreline in more recent times. Sediment cores collected offshore of Galveston Island also record its growth in the form of sand layers that were deposited near shore within the shoreface (fig. 1.2). These sand layers are now buried beneath a layer of mud that contains shells of animals that live seaward of the shoreface. As the island began to retreat landward, marine mud was deposited on top of the sand. Radiocarbon ages of shells within this mud show that this change from growth to erosion took place around 1,200 years ago. Hence, the island stopped growing about 1,200 years ago and since then has been retreating landward.

Is the current rate of retreat unprecedented?

The dashed line through the water column in figure 2.6 is the best estimate of the shoreline location 1,200 years ago, assuming an offshore profile similar to that of today. Using this shoreline reconstruction, it is possible to estimate a rate of retreat of about 3 feet per year for the past 1,200 years, which is about the rate at which the shoreline is currently eroding at this location. Thus erosion predates human influence such as the construction of dams, the Galveston jetties, and groins. But current rates of erosion do exceed 3 feet per year in a number of locations and can, in most cases, be attributed to human influence (see chapter 5). It is noteworthy that the shoreline that retreated across the shelf as sea level rose has left hardly a trace of its existence. This reflects the capacity of

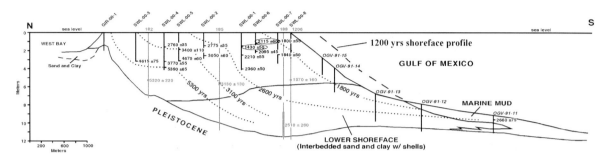

Figure 2.6. Using Donax shells collected within sediment cores through Galveston Island, we have been able to reconstruct the evolution of the island in detail. When this work was done, from 2000 through 2003, the radiocarbon methodology had improved enough to allow dating of small samples, such as a single Donax shell, with accuracy to within 100 years. The dotted lines are time lines through the island, showing the width of the island at these times. Modified from Rodriguez et al. 2004.

transgressive ravinement, the process whereby waves cut deep into the coastal succession as the shoreline retreats landward, to completely remove coastal barriers as the shoreline retreats landward (chapter 1).

Bolivar Peninsula's Evolution

About 5,000 years ago, the Sabine Bank shoreline coexisted with the oldest part of Galveston Island, and the shoreline was far more sinuous

Aerial view of Bolivar Peninsula. Photo from GlobeXplorer.

than it is today (fig. 2.3). About 2,500 years ago, the shoreline shifted landward, reaching its present position only a few centuries later. This is one of the fastest episodes of coastal change recorded along the Texas coast. While Sabine Bank was still a barrier island, a vast bay, ancestral East Bay, existed between the bank and the present shoreline. Sediment cores collected offshore of Bolivar Peninsula have sampled these old bay deposits, which contain abundant oyster shells that are currently being exhumed by waves and transported onto beaches between High Island and Bolivar. Resting above the oyster shell-bearing bay mud is mud that contains fossils of mollusks that live only in the offshore waters of the Gulf. The contact between these sediment layers records the relatively short interval of time during which the old Sabine shoreline was drowned and the bay joined with the Gulf. As the Gulf waters encroached farther landward, a spit of land began to form along the east end of what is now East Bay. This was the first step in the evolution of Bolivar Peninsula.

Bolivar Peninsula formed in a different manner from Galveston Island. The peninsula grew by what geologists refer to as *spit accretion*, the process by which coastal currents transport sand along the coast to form a spit. With time, the spit grew toward the west, under the influence of longshore transport, and became Bolivar Peninsula.

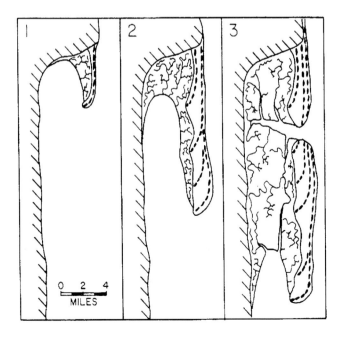

Figure 2.7. Model for development of a barrier island from a spit. The dashed lines represent beach ridges. Modified from Hoyt 1968.

Compared to Galveston Island, Bolivar Peninsula is a relatively young feature. It began to form after 2,000 years ago, and most of it formed within a few centuries. The development of the peninsula is recorded in sand pits and in sediment cores from onshore and within East Bay. The photograph in figure 2.8 was taken in one of several sand pits on the peninsula and shows clean sand, which contains shells of *Donax*, resting on marsh and lagoon deposits that are mostly mud with oyster shells. These units are separated by a shell bed that marks the initial emergence of the barrier. Radiocarbon dates from *Donax* shells indicate that the barrier began to develop at this location about 1,500 years ago.

Bolivar Peninsula grew westward through a process called *lateral accretion,* the addition of sand, layer by layer, to the westward end of the peninsula (figs 2.7 and 2.9). The rate of westward growth slowed as the peninsula advanced over the old Trinity River valley because of the greater amounts of sediment needed to fill the valley. Seismic profiles and long sediment cores show that the thickness of sand composing the peninsula increases toward the west, reaching a maximum thickness of 60 feet beneath the town of Port Bolivar.

Two distinct sets of beach ridges occur on Bolivar Peninsula; an older

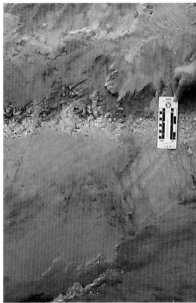

Figure 2.8. In this sand pit on Bolivar Peninsula, sand rests on clay. The sand contains Donax shells, and the clay contains oyster shells. Hence, the change represents barrier sand that migrated landward onto bay mud. On the right is a close-up of a shell bed at the base of the beach ridge deposits.

West   East

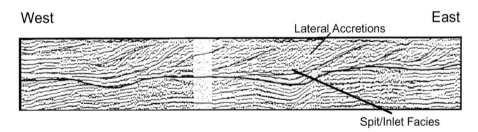

Lateral Accretions

Spit/Inlet Facies

Figure 2.9. The west-ward growth of Bolivar Peninsula is recorded in a seismic line collected along the Intracoastal Waterway. Think of the seismic reflectors shown in this figure as sand beds that through time are stacked upon one another, filling the flanks of the ancestral Trinity River valley, which lay to the west of the evolving peninsula. This type of alongshore growth is called lateral accretion. The profile is one mile long.

set on the back side marks an earlier phase of barrier evolution. These ridges curve toward the north, reflecting the curvature of the old shoreline at the west end of the barrier (figs 2.7 and 2.10). Seaward of the curved ridges is a younger set of ridges that trend parallel to the modern shore-line. The two ridge sets are separated by a definitive boundary, which is highlighted in figure 2.10 with a dashed line. Radiocarbon dates from *Donax* shells in the older beach ridge set range from 1,500 to 1,200 years at the west end of the peninsula. *Donax* shells from the younger set of beach ridges on Bolivar Peninsula have yielded radiocarbon dates younger than 800 BP. These data indicate that around 800 years ago, the peninsula was almost decapitated by a major storm or series of storms. This event

Figure 2.10. Aerial view of two sets of beach ridges on Bolivar Peninsula, a younger set that trends parallel to the shoreline and an older set on the back side of the peninsula that curves toward East Bay. The dashed line marks the boundary between these two sets.

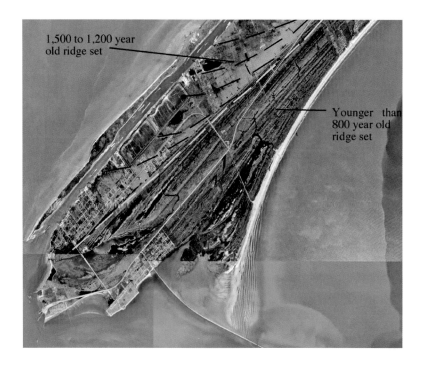

1,500 to 1,200 year old ridge set

Younger than 800 year old ridge set

was followed by an episode of rapid growth of the peninsula. At no time since then does the peninsula appear to have been so severely impacted by a hurricane with such magnitude.

Follets Island's Evolution

Follets Island, located west of Galveston Island, actually is no longer an island, but when it was first mapped it was apparently not connected to the mainland at its western end. A relatively low barrier island, with elevations mostly less than 4 feet, it has been completely submerged by tropical storms, such as Hurricane Carla, in historical time.

Follets Island sits atop red clay deposited by the Brazos River when it flowed through this area in what is now Bastrop Bayou. Sediment cores through the barrier sampled between 3 and 8 feet of barrier sand, so the island is quite thin relative to Galveston Island and Bolivar Peninsula. Unlike Galveston Island, Follets Island is narrow, and a significant amount of the sand that is being eroded from the island washes over into back-barrier wetlands. This stage of barrier island evolution, referred to as the "rollover" stage, is the final stage before the island is completely eroded.

Aerial view of Follets Island and Christmas Bay, which is located on the landward side of the island. Photo from GlobeXplorer.

Aerial view to the north across Follets Island. County Road 3005 follows the narrow boundary between the beach and back-barrier wetlands.

The island is highly prone to breaching during storms, as evidenced by numerous storm channels and wash-over deposits on the landward side of the island.

Surfside Area

West of Follets Island, the east Texas barrier system ends and the extensive Brazos and Colorado delta plain begins. In Surfside, large dunes, a rarity along coasts to the east, indicate that there is a significant source of sand nourishing the beaches. The sand is being derived from an ancestral channel of the Brazos River, Oyster Creek. As recently as 1,500 years ago, the Brazos River occupied this channel and flowed into the Gulf just east of Surfside. At the time the river was flowing through this area, the shore-

Aerial view of coast between the west end of Follets Island and the mouth of the Brazos River. From the U.S. Geological Survey.

line was some distance offshore, so the river valley extended offshore. Waves have reworked the sands from the valley, and those sands were delivered back into the longshore transport system. Still, Surfside Beach has a recent history of rapid erosion. This erosion is the result of human intervention, specifically the alteration in the course of the modern Brazos River from its original channel, which extended through Freeport, to a location approximately 6 miles to the west (see chapter 3).

The Brazos Alluvial Plain

The Brazos River delivers more sediment to the Texas coast than any other river. During the last 15,000 years, while sea level was rising rapidly, the Brazos transported so much sediment that it filled its deep river valley with sediment as fast as sea level rose. Hence, it was never flooded to create a bay. In fact, the river filled the valley that was cut when sea

Aerial view of the Surfside Beach area showing Oyster Creek as it meanders to the coast. Oyster Creek occupies an old channel of the Brazos River, which was in this location between 4,000 and 1,500 years ago. The river then shifted its course to its pre-1929 location, which flows through Freeport into the Gulf at the Freeport jetties. Sediment cores collected offshore of Oyster Creek reveal that the channel has been completely exhumed by the advancing shoreline. From the U.S. Geological Survey.

level was at a low stand and shifted to three separate channels, filling them all with sediment as sea level continued to rise. These combined ancestral river channels now comprise an extensive area that is referred to as the Brazos alluvial plain.

The youngest river channels of the Brazos River can still be seen in aerial photographs of the region, and using radiocarbon dates the ages of the individual valleys have been determined. Between approximately 6,000 and 4,000 years ago, the Brazos River flowed through what is now Bastrop Bayou, north of Christmas Bay. This is referred to as the Big Slough channel. About 4,000 years ago the Big Slough channel was abandoned when the river shifted west to occupy the Oyster Creek channel. It remained in that location until 1,500 years ago. At that time the river again shifted its course to the west, where it initially flowed through what is now Jones Creek. It later shifted to its modern channel near Freeport. It remained there until 1929, when the U.S. Army Corps of Engineers diverted the lower part of the river to a location 6 miles west of Freeport.

Figure 2.11. The locations of ancestral Brazos River channels. Courtesy of Patrick Taha.

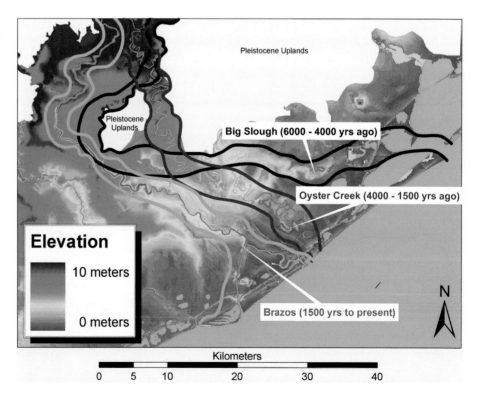

This natural shifting of river courses, which geologists refer to as avulsion, is what creates alluvial plains. The Colorado River had a similar history of avulsion, and together the Brazos and Colorado rivers formed an alluvial plain that occupies the coast from Follets Island to Matagorda Bay.

The Origin of East Texas Bays

Coastal rivers are strongly influenced by sea level change. This is because they cut their valleys to the level of the sea. If sea level falls, rivers cut deeper valleys. When sea level rises, the rivers either fill their valleys with sediment or flood to become bays. During the last glaciation, when sea level was approximately 350 feet below present, the rivers of east Texas cut valleys that were approximately 100 feet deeper than the modern river valleys where they reach the present shoreline. The ancestral Trinity and San Jacinto rivers merged near the center of modern Galveston Bay, and this valley extends offshore to the edge of the continental shelf.

During the rise in sea level of the past 17,000 years, rivers that were transporting large amounts of sediment, such as the Brazos and Colorado, began to backfill their valleys with sediment. These are the rivers

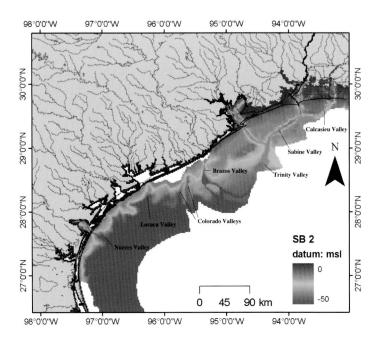

Figure 2.12. This digital elevation map of the land surface 20,000 years ago highlights the incised Calcasieu, Sabine, Trinity, Brazos, and Colorado river valleys. As sea level rose during the last several thousand years, the Calcasieu, Sabine, and Trinity valleys were flooded to create Calcasieu Lake, Sabine Lake, and Galveston Bay respectively. Courtesy of K. Milliken.

that currently flow into the Gulf. The advancing sea overcame smaller rivers with less sediment supply, like the Trinity, Sabine, and Calcasieu rivers, flooding their valleys to create bays. Likewise, Matagorda Bay is the drowned river valley of the ancestral Lavaca, Palacious, and Karankawa rivers, and Corpus Christi Bay is the drowned Nueces River valley. With time the bays are being filled with sediment that will record the overall rise in sea level and flooding of the valley to create the bays.

Sabine Lake's Evolution

Sabine Lake is the drowned Sabine River valley (fig. 2.12). Detailed study of the lake has shown that it was first flooded about 8,000 years ago, when sea level was rising rapidly. For much of its history, Sabine Lake was occupied by a large bayhead delta, which stepped landward to its present position at the head of the bay only about 2,000 years ago. This landward shift occurred when sea level was rising slowly, perhaps as a result of a decrease in sediment supply to the lake. At about that

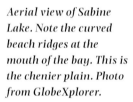

Aerial view of Sabine Lake. Note the curved beach ridges at the mouth of the bay. This is the chenier plain. Photo from GlobeXplorer.

same time, the lake was shallow enough that its mouth was filled by sand ridges, which together comprise what is called a *chenier plain.* The curvature of these ridges reflects the infilling of the mouth of the lake through seaward growth. Radiocarbon age dating of these ridges has shown that they formed after about 2,500 years ago.

Galveston Bay's Evolution

Galveston Bay is the flooded river valley of the Trinity and San Jacinto rivers (fig. 2.12). Using seismic data and sediment cores collected within the bay, geologists have been able to reconstruct the bay's evolution with considerable detail. These data show that Galveston Bay has had a history of rapid and dramatic change, in response to sea level rise across the irregular topography of the old river valleys and changes in climate that regulated sediment supply to the bay. During the lowstand in sea level of approximately 18,000 years ago, when sea level was at its lowest point and the shoreline was situated many tens of miles seaward of its current

Aerial view of Galveston Bay. Photo from GlobeXplorer.

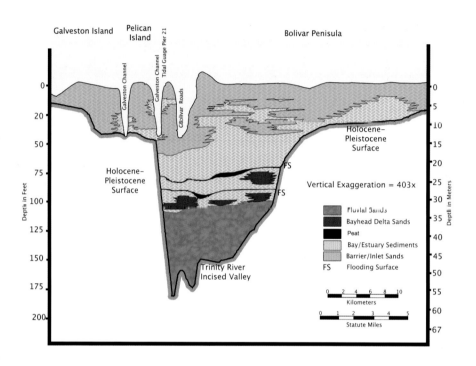

Figure 2.13. This geo-logic cross section of the old Trinity River valley was constructed from seismic data and long sediment cores collected along the Intracoastal Waterway, between Galveston Island and Rollover Pass at the east end of Bolivar Penin-sula. It illustrates the manner in which the sediments that fill the valley record Galves-ton Bay's evolution as well as the evolution of Bolivar Peninsula and Galveston Island. Note also the shape of the val-ley and imagine what it looked like as this valley was gradually filled with water to create the bay. Modified from Siringan and Anderson 1993.

position, the Trinity River cut a valley about 170 feet deep where it flowed through the area that is now Bolivar Roads. As sea level rose after the ice sheets began to melt, the river initially filled the valley with sand and gravel. Approximately 10,000 years ago, the rate of sea level rise exceeded the river's capacity to fill its valley with sediment, and the valley was flooded to create a narrow bay, similar to present-day Chesapeake Bay. As sea level continued to rise, the broader flanks of the valley were flooded, and the bay took on a more rounded shape, similar to the present bay.

The history of Galveston Bay's evolution is best recorded in a series of long drill cores collected along the axis of the bay. These cores record the initial flooding of the river valley to form a delta at the head of the bay (bayhead delta). The delta overlies the old river deposits. The delta de-posits consist of mud and sand with abundant organic material, similar to deposits in the extensive wetlands of the modern Trinity delta. These deposits contain shells of *Rangia,* a mollusk that today lives in the shallow waters surrounding the Trinity delta (fig. 2.14). Resting above the *Rangia*-bearing bayhead delta deposits of Galveston Bay is mud with oyster shells and other fossils that indicate a bay setting similar to the present Galveston Bay environment.

Figure 2.15 illustrates how the major bay environments, which include central basin (open bay) mud, bay-head delta, and flood-tidal delta environments, have stepped landward through time as the bay was flooded by rising sea level. Similar sequences to those sampled in Galveston Bay have been sampled in Sabine Lake, Matagorda Bay, and Corpus Christi Bay. In all of these bays, the contact between the different sediment layers and the environments they represent are often quite sharp, which implies that there were times when these bays changed dramatically. But how rapid were these changes and what caused them? Will similar changes occur in the future in response to changing rates of sea level rise and climate change? To get at this question, we need to take a closer look at the sediments that fill the bay and gather radiocarbon ages that enable us to measure rates of change in the past. Using these data, we are able to reconstruct the stages of evolution of Galveston Bay and other bays in the region in considerable detail.

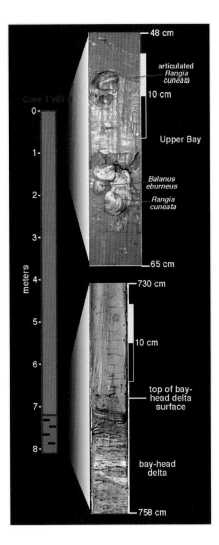

Figure 2.14. This example of a sediment core from Galveston Bay records a marked change in the bay setting. At the base of the core are sediments that have abundant plant debris but lack shells. This sediment type occurs today on the surface of the Trinity bayhead delta, or delta plain. There is a distinct change at about 740 cm in the core, above which greenish mud with abundant shells exist. The shells are those of Rangia, a clam that today lives only in the upper reaches of the bay, offshore of the delta. Thus, in this case, the change represents a flooding event when the subaerial delta plain was drowned.

About 8,200 years ago, ancestral Galveston Bay was situated seaward of the modern bay, and extended about 30 miles offshore to what is now Heald Bank (fig. 2.3). A rapid flooding event occurred 7,700 years ago. During that event, the Trinity bayhead delta shifted landward 12 miles within a couple of centuries, causing major reorganizations of bay environments. As a result of this flooding event the bay area was nearly doubled in a few centuries. During this time interval, rates of migration along

Much of the sediment being transported by the Trinity River to Galveston Bay is deposited in the Trinity bayhead delta. The area of the delta that is above water is the delta plain. Photo from GlobeXplorer.

the bay shore averaged 50 feet per year, and the bay was choked with organic material derived from the inundation of wetlands that lay adjacent to it. Until 5,500 years ago, Galveston Island did not exist, and Bolivar Peninsula did not form until after 2,500 years ago. So the bay mouth was wide, and the saltier Gulf waters had greater influence.

Are the changes we observe during our lifetime unprecedented?

Figure 2.16 summarizes the stages of evolution of the upper Texas coast. At around 15,000 years ago, the ice sheets were melting and sea level was rising rapidly. At that time, the shoreline was still situated nearly 80 miles seaward of its present location and a chain of coral reefs, including the Flower Gardens, extended offshore along the coast. The city of Galveston would have been located on the banks of the Trinity River, whose valley was over 170 feet deep at that location.

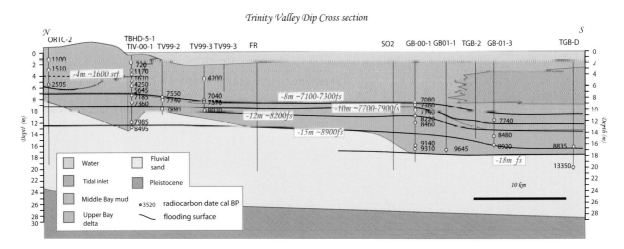

Figure 2.15. Sediment cores from Galveston Bay were used to study the evolution of the bay and reveal that at times the bay underwent rapid change. This cross section down the axis of the bay was constructed from sediment cores that record changes in bay environments as the Trinity and San Jacinto river valleys were flooded by rising sea level. Modified from Rodriguez et al. 2005.

Eight thousand years ago, the Gulf shoreline was situated about 30 miles (approximately 50 kilometers) seaward of its present location, where Heald Bank is now located (fig. 2.3). Galveston Bay was a long, narrow bay that extended north to the present location of Smith Point and San Leon. A dramatic landward shift in the shoreline took place 7,700 years ago, forming the upper part of Galveston Bay. By 5,500 years ago the Gulf shoreline had shifted to a location near Sabine Bank. West of the Trinity River valley, the shoreline curved to the north to a new location somewhere within what is now West Bay. Flooding of the low-lying areas around Galveston Bay resulted in its present rounded shape. At this time, the Brazos River flowed through what is now San Luis Pass and nourished a delta, the Big Slough delta (fig. 2.11). Remnants of this delta still exist offshore of the pass.

Two thousand years ago, Galveston Island had grown to about two-thirds its current size, and Bolivar Peninsula was just beginning to form. The Brazos River flowed into the Gulf at Surfside Beach, where it constructed a new delta. By 1,200 years ago the coastal setting was similar to today, and the current phase of landward retreat of the shoreline began.

Figure 2.17 is a map of the main geological components of the upper Texas coast. Now that we understand how these different components evolved, we are ready to consider historical changes in our coast. But we need to keep in mind that the natural evolution of our coast has been punctuated by events that were quite rapid, such as the rapid advance of

Figure 2.16. This set of block diagrams summarizes the evolution of the upper Texas coast. Note that elevations are highly exaggerated. Courtesy of K. Milliken.

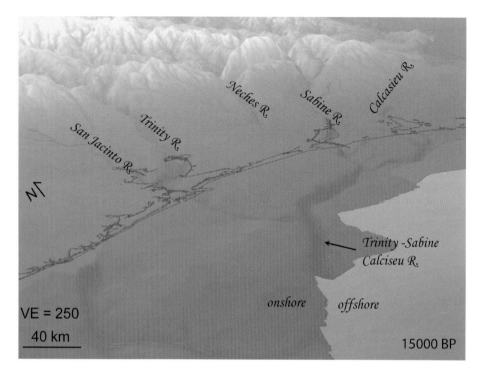

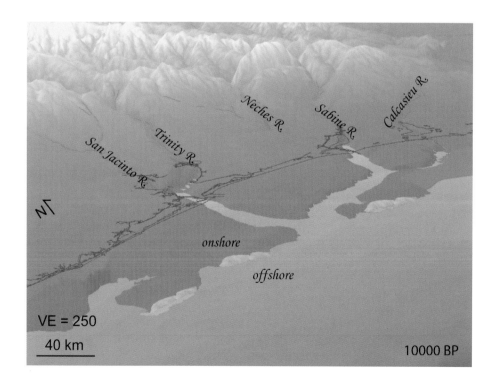

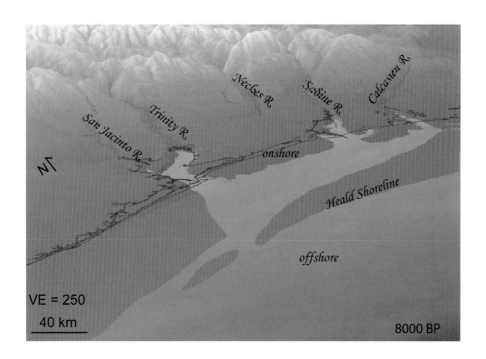

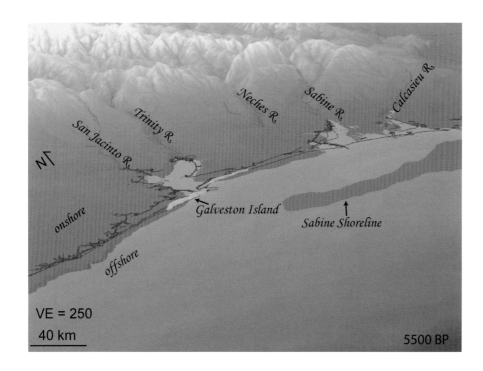

Figure 2.17. The major geological components of the upper Texas coast. Note that the Sabine/ Neches and Trinity/San Jacinto river valleys have been flooded to create Sabine Lake and Galveston Bay, respectively. The sediment supply of these rivers was not high enough to keep pace with the rate of rising sea level, hence these valleys were flooded to create these bays. In contrast, the Brazos and Colorado rivers, which have much larger drainage basins and therefore larger sediment supplies relative to the Sabine, Neches, Trinity, and San Jacinto rivers, filled their valleys as sea level rose at the end of the last glaciation. Hence, they now occupy a vast alluvial plain that dominates the coast for nearly 60 miles. Modified from Bernard et al. 1970.

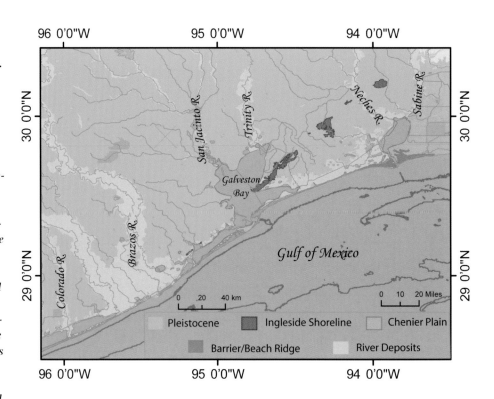

the shoreline and associated drowning of barrier islands to form banks and the dramatic changes in Galveston Bay that occurred during its early evolution. For the most part, these changes occurred during the time when sea level was rising faster than today. Other changes resulted from variations in sediment supply to bays and the coast, and these changes were caused by climatic events, specifically variations in precipitation. To date these changes have remained unprecedented in historical time, but that could change this century.

3

Historical Changes in Our Coast

The Galveston Seawall is a last line of defense against the encroaching shoreline.

Are humans responsible for changes in our bays and shorelines?

In Victorian times, Galveston Island was the "Wall Street of the South," a boomtown whose growth was largely spurred by its warm climate and proximity to the Gulf. This was before the Houston Ship Channel, Galveston jetties, and Texas City Dike were constructed. At that time, there was a large sandbar (ebb tidal delta) offshore of Bolivar Roads, the tidal inlet between Galveston Island and Bolivar Peninsula. This bar offered a formidable challenge to sailing vessels coming and going to the port of Galveston. But there was a deep tidal channel that extended from the Gulf to the back side of the island, and that is why Galveston was a thriving port city at the turn of the last century.

In the early fall of 1900, Galveston Island was struck by a powerful storm that virtually destroyed the city and killed most of the island's inhabitants. The city was rebuilt and again thrived in the early part of the century. Construction of the Galveston Seawall began in 1902, and not long after the turn of the century, dredging of the Houston Ship Channel began, with the development of steam-powered dredging. The new ship channel enabled ships to transport their products farther inland, closer to major rail lines and highways. That was a serious blow to Galveston's economy, and the island fell on hard times. Today the island again enjoys a measure of prosperity, kindled largely by tourism. The island's historical district gives tourists a true feeling of life in Victorian times. Other attractions, like Moody Gardens and the Lone Star Flight Museum, provide educational opportunities for vacationing families. But it is probably safe to say that the beaches are the main attraction, and without them, tourism and the island's economy would suffer. Unfortunately, the beaches are slowly disappearing.

The Houston Ship Channel

Ironically, the dredging of the Houston Ship Channel has had a second impact on the city of Galveston, one that is only now becoming evident. During the early years of channel maintenance, much of the dredge

This turn of the twentieth century photograph was taken shortly after the Galveston Seawall was constructed. Note how wide the beach was at this location. Courtesy of Rosenberg Library.

West Bay

Texas City dike

East Bay

Figure 3.1. This satellite image of the lower portion of Galveston Bay shows sediment-laden water that is flowing along the west shore of the bay being diverted by the Texas City Dike so that it flows directly into the Gulf and not into West Bay.

spoil from the bay was hauled offshore and dumped near the jetties, atop the old ebb tidal delta. The tidal delta would be a viable sand resource for beach nourishment, except that it is largely covered by contaminated, muddy dredge spoil.

The sediment dredged from the Houston, Texas City, and Galveston ship channels was also piled along the sides of the ship channel and into artificial islands, such as Pelican Island. Prior to the dredging, the island was a natural feature, part of the flood tidal delta, and undoubtedly an important bird sanctuary at the turn of the century. The dredge spoil and ship channel extend the entire length of Galveston Bay and have exerted a significant impact on it. During dry spells, when freshwater input to the bay is minimal, salt water from the Gulf flows far up the bay as a wedge of salt water, or "salt wedge." Yet the bay's ecosystem has in large part adjusted to these changes, and Galveston Bay is, relatively speaking, still a healthy estuary. Indeed, some of the changes man has made may have had a beneficial effect on the bay system. The Texas City Dike, for example, virtually blocks the east end of West Bay, preventing sediment-laden water and associated pollutants that move along the west shore of Galveston Bay from entering West Bay. This has significantly decreased sedimentation in West Bay. The bay is clearer than in the past and has been somewhat protected from the influx of contaminated waters from Galveston Bay, especially in the postwar era when pollution in Galveston Bay was unchecked.

Groundwater Withdrawal

One of the most dramatic impacts humans have had on the upper Texas coast resulted from substantial withdrawal of groundwater from aquifers beneath the more industrialized area of the Houston Ship Channel. This caused 4 to 7.5 feet of land subsidence in the area immediately between Houston and Baytown within a few decades. A smaller area of subsidence exists near Texas City. The entire eastern shoreline of Galveston Bay was subjected to the impacts of subsidence, which accelerated coastal erosion and inundation of low-lying areas. It is estimated that more than 50,000 acres of wetlands in Galveston Bay were lost during the time subsidence was at its peak.

In the mid-1970s, the Harris/Galveston Coastal Subsidence District was created to limit groundwater use in the region. But the practice of groundwater usage for industrial purposes continues in the Freeport area, and it has resulted in up to 2.5 feet of subsidence, which has contributed to the high rates of beach erosion that occurs at Surfside (http://texascoastgeology.com).

Figure 3.2. Land surface subsidence in the Houston-Galveston area between 1943 and 1973. From Gabrysch and Bonnet 1975.

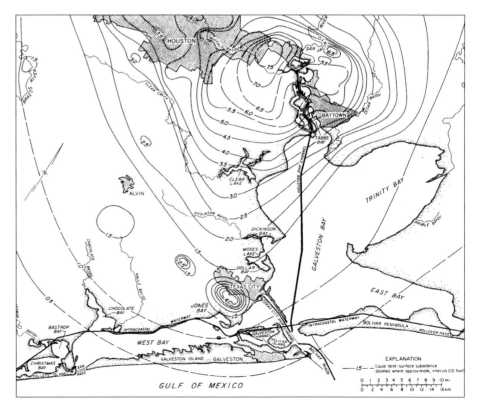

This 1975 photograph of the Baytown area shows the highway that was submerged as a result of land subsidence.

Damming Rivers

All of the rivers that empty into the upper Texas coast have been dammed. Some have argued that damming rivers prevents them from delivering sand to the coast and causes our beaches to erode. This is not entirely true. In fact, rising sea level flooded most rivers to create bays, and the sand that these rivers deliver to the coast is now being deposited in the upper portions of these bays in deltas, such as the Trinity Delta at the head of Galveston Bay. Damming of the San Jacinto and Trinity rivers has, in fact, reduced their sand supplies, and while the impact on the coast has been insignificant, there is evidence that the damming of the Trinity River has impacted the wetlands of the upper Trinity Bay area. Specifically, the Trinity delta plain shows signs of shrinking in recent years (White, Morton, and Holmes 2002). Silt and clay that is suspended in the rivers is not as effectively trapped by the dams. As a result, the relative amounts of sand versus silt and clay has been changed. This has had an adverse effect on the wetlands by inundating them with fine sediment while at the same time reducing the sand that is needed as a framework for wetlands growth.

The modern Trinity River Delta began its growth about 2,000 years

Figure 3.3. The pho-
tograph shows the
modern Trinity Delta.
The dashed line shows
the outline of the delta
based on an 1852 map.
By the early 1900s, the
delta had grown to the
east and merged with the
eastern shore of Trinity
Bay, blocking the upper
reaches of the bay to
form Lake Anahuac,
previously Turtle Bay.
Photo from the U.S.
Geological Survey.

ago, following the most recent flooding event in Galveston Bay. The delta
has grown southward in historical time to block the upper part of Trinity Bay and form Lake Anahuac. Following the Civil War, Wallisville,
which is located upstream of the delta on the banks of the Trinity River,
was a thriving community with an active lumbering and shipbuilding
industry. The city and its industries were virtually destroyed by a hurricane in 1915, although shipbuilding continued until after World War
II. During this time, the Trinity River was used as a navigation channel
between Wallisville, and other towns along the river, and Galveston Bay.
Until this time the sediment delivered by the Trinity River to the delta
was dispersed through several distributary channels that nourished the
entire delta with sediment. This natural process of delta growth ceased
when the river was converted to a navigation channel, and sediment was
forced to move through the river mouth into the bay. The result has been
that the river mouth has extended its channels farther into the bay, and

much of the sediment that previously was delivered to the delta plain is now deposited in Galveston Bay.

The Brazos and San Bernard rivers are the only rivers that today flow into the upper Texas coast and deliver sand to the coast. However, the amount of sediment these rivers deliver has decreased significantly in the last few thousand years because rising sea level has reduced the grade of the rivers and therefore their capacity for sand transport. Dam construction has compounded this natural effect and diminished sediment delivery to the coast.

The rate at which sand moves downriver toward the coast is measured in decades to centuries, so the full impact of dam construction has yet to be felt at the coast. Over the three decades that I have been taking students to the Brazos and Trinity rivers, I have watched the sandbars along

Aerial view of sandbars along the lower Brazos River.

these rivers shrink. I have also observed the significant impact floods have in transporting sand down these rivers in a very episodic fashion.

Of those rivers that flow into the upper Texas coast, only the Brazos River is a significant supplier of sand to the coast today. It provides a spectacular example of the impact man can have on the coast and the rate at which human impact is felt. In 1929 the U.S. Army Corps of Engineers changed the course of the Brazos River by dredging a new channel that extended from the city of Freeport to a location some 6 miles west of the Freeport jetties, where the river once flowed into the Gulf. The objective was to convert the old river mouth into a ship channel and prevent sedimentation of the channel. That was a success, but not without a price.

The diversion of the Brazos River virtually eliminated the sand supply to Surfside Beach and resulted in initial erosion rates of between 50 and 100 feet per year, culminating in the virtual destruction of the delta (fig. 3.4). Meanwhile, sand that was removed from the old delta was trans-

Figure 3.4. These photographs illustrate changes in the coast associated with the diversion of the Brazos River near Freeport in 1929. The photograph at the bottom was taken in 1930, just after the river was diverted; the photograph at the top was taken in 1989, 60 years after the river diversion.

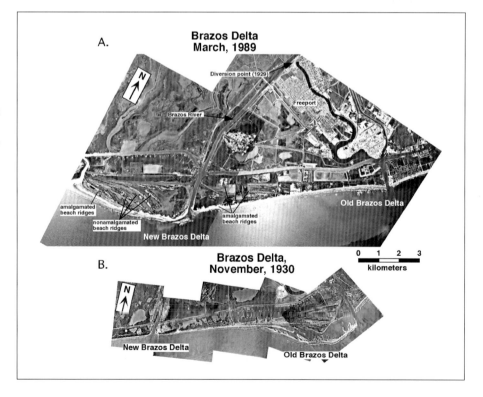

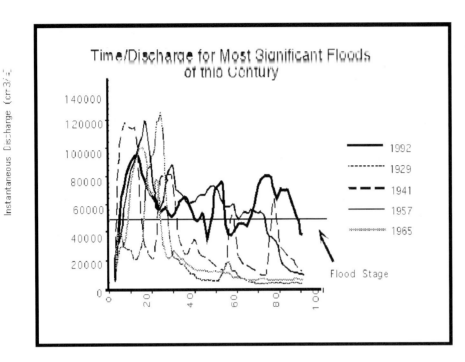

Figure 3.5. The Brazos River experiences a major flood about every 10 to 15 years, during El Niño years. Each time the river floods it delivers a new supply of sand to the upper Texas coast.

ported to the west by longshore currents and accreted to the new delta. Most of the sand relocation occurred within thirty years of the river diversion. This is a good illustration of the efficiency of longshore transport.

Recent studies have shown that the supply of sand to the coast by the Brazos River is highly episodic and is controlled by climatic events that operate on a decadal time scale. The river floods about every ten to fifteen years. These floods occur during El Niño years, when significant amounts of moisture are delivered to Texas from the West Coast. During these floods, the Brazos River delivers large quantities of sand to the coast. The most recent flood occurred in 1992 when the river was flooded for nearly three months. Following the 1992 flood, enough sand was delivered to the coast to create a small island offshore of the river mouth. Within a year after the 1992 flood, most of the sand that had been delivered to the coast had been transported back onshore and accreted to the shoreline to form a prominent beach ridge. The longshore current transported the remaining sand to the west. Very little sand was transported to the east toward Surfside Beach. This is another excellent example of the rate at which waves and longshore currents redistribute sediments.

Aerial photograph of the Brazos Delta taken ten months after the 1992 flood. Note the small island off the river mouth and the prominent ridges and swells located landward of the beach. The ridges were formed from bars created during prior floods and were later moved onshore by waves and accreted to the delta. Therefore each ridge marks a past flood event and associated pulse of sediment to the delta. If you walk along these old ridges, you can find old bottles with dates that record major floods of this century (figure 3.5).

Given the episodic manner in which sand is delivered to the coast by the Brazos River, the full impact of damming the river has not yet been felt. But, it is safe to say that future generations will see the impact in the form of accelerated coastal erosion around and west of the delta.

Sea Level Rise and the Destiny of Our Coast

The continuous landward movement of the shoreline has claimed many houses along the upper Texas coast.

Why is sea level rising and how will the rate of sea level rise vary in the future?

Warming Climate and Rising Sea Level

For the most part, the changes that are occurring along our coast, such as coastal retreat and wetlands loss, are due to rising sea level and a shortage of sediment supply to the coast. Sea level rise has two components. One is the actual rise in the ocean's surface, which is called *eustasy*. The other is subsidence of the land surface. The sum of these two is relative sea level rise. Let's look first at eustasy.

Global sea level is currently rising relatively slowly, at a rate of between 1.5 and 2.0 millimeters per year (approximately 1/16 to 1/8 inches per year), and has been rising at this relatively slow rate for about the past 5,000 years (fig. 2.2). The rise is caused by melting of glaciers and ice sheets and by warming of ocean waters, which causes them to expand. Scientists predict that both processes will increase in the future because of global warming (IPCC 2001).

While debate continues over global warming, the historical increase in Earth's surface temperature during the past century is irrefutable. Most of the debate about global warming focuses on the causes of this increase, specifically how much is due to greenhouse gas emissions into the atmosphere and how much may be part of a natural climate cycle. We know that Earth's temperatures have varied in the past and that temperature increases have been correlated to natural increases in the concentration of greenhouse gases. The opponents of global warming point to these changes and argue that they have happened in the past and without human influence. Regardless of the cause, the Earth is getting warmer, glaciers and ice sheets are shrinking, and the rate of relative sea level rise is increasing. We may as well accept these facts and get on with the business of planning for a warmer future and the changes it will bring.

Those who study the changes that will occur as the Earth's atmosphere continues to warm have argued that the initial and most profound changes will occur in the polar regions, where temperature changes are most extreme. Their predictions are proving correct, as the polar latitudes are experiencing unprecedented warming trends. Arctic sea ice cover has been decreasing since the 1970s, with the least sea ice cover on record occurring the past three summers. In the Antarctic Peninsula, atmospheric

Figure 4.1. The overall increase in Earth's surface temperatures has varied, but there is little question that temperatures have been slowly rising. From IPCC 2001.

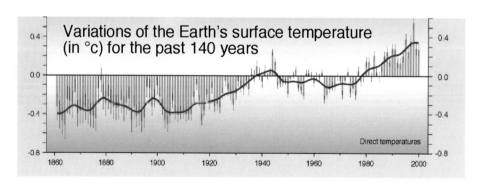

temperatures have increased nearly 2°C in the past century, and significant and unprecedented retreat of ice shelves in the region has occurred. The most spectacular event was the breakup of the northern portion of the Larsen Ice Shelf in 2002. That event inspired a movie, *The Day after Tomorrow*. Like most natural disaster movies, this one was far from reality. However, it did raise public awareness of the fact that sea level is controlled by factors that are beyond our control. The fact is, there is potential for rapid sea level change, and the impact on coasts would be profound, but the large cities of the world would not be flooded. Data acquired in the past few years tell us that the margins of the West Antarctic Ice Sheet are beginning to shrink and that the volumes of meltwater added to the oceans is contributing to global sea level rise. While the changes that have occurred in the Antarctic Peninsula have not significantly influenced sea level, they may herald a warming trend that could cause sea level to rise faster in the future.

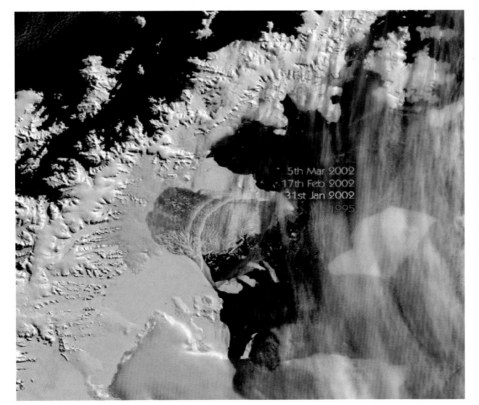

Figure 4.2. Temperatures in the Antarctic Peninsula increased about 2°C during the past century, resulting in what are believed to be unprecedented changes in the region. The most notable change was the breakup of the northern portion of the Larsen Ice Shelf in 2002, shown in this satellite photograph. Courtesy of the British Antarctic Survey.

Modern coasts and estuaries evolved during the interval of relatively slow sea level rise that began approximately 5,000 years ago (fig. 2.2). Prior to that time, sea level was rising at a much faster rate, in response to ice sheet melting in both hemispheres. It was probably not long after Galveston Island was formed that the first people settled there, the ancestors of the Karankawa. Generation after generation, they watched the island slowly evolve. These early coastal inhabitants were conditioned to rapidly advancing shorelines, as much as 50 feet per year, and the immense damage inflicted by storms. They were also living along the margins of bays where dramatic changes in environment occurred within a century or two (see chapter 2).

During the previous interglacial period, approximately 120,000 years ago, sea level was about 15 feet higher than present, and the shoreline was located approximately midway up Galveston Bay at the present location of Smith Point and San Leon Point. The next time you drive south to Galveston Island, look to the east as you drive through the town of Dickinson. Notice the tall pine trees that live atop the old shoreline within the thick sandy soils that formed along this relatively narrow but extensive shore. Geologists refer to this old shoreline as the Ingleside shoreline because it is best preserved near the town of Ingleside on the east bank of Corpus Christi Bay (fig. 2.17).

Why was sea level 15 feet higher during the previous interglacial?

The answer to that question is a bit disconcerting. If the "unstable" portions of the current Antarctic and Greenland ice sheets were to melt, sea level would rise about 15 feet. So we may still be moving into the maximum interglacial state. If 15 feet sounds unrealistic, consider the fact that if the entire Antarctic Ice Sheet were to melt, sea level would rise nearly 200 feet. For the most part, the ice sheet is and has been stable for many thousands of years, so we can rest easy. But parts of the ice sheet are grounded far below sea level, and these are what glaciologists consider to be the unstable portions. Recent studies have shown that the margins of the West Antarctic Ice Sheet are getting thinner. This is not good news.

The Intergovernmental Panel on Climate Change (IPCC) has published predictions for global sea level rise over the next century. Scientists are still debating the reliability of these predictions, but the debate focuses

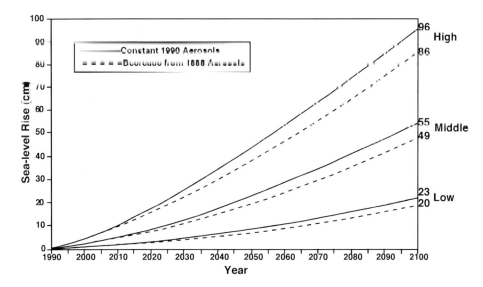

Figure 4.3. The predicted sea level rise for the next century. From IPCC 2001.

on the magnitude of sea level rise; there is little question that sea level will continue to rise in the next few centuries. The greatest uncertainty in these predictions is how the Antarctic Ice Sheet will respond to global warming. If the ice sheet does contribute, we could experience almost a meter rise over the next century.

The most recent findings on global sea level rise come from satellite altimetry data. These data indicate the average rate of sea level rise has increased from 1.8 ± 0.3 mm/yr to 3.0 mm/yr during the past 12 years (Cazenave 2006).

Coastal Subsidence

The combined global rise in sea level due to ocean water expansion and ice sheet melting (eustasy) and subsidence is referred to as relative sea level rise. It is the relative rise of sea level that concerns us most because the Louisiana and east Texas coasts are the most rapidly subsiding coasts in the United States. In fact, the rate of relative sea level rise in Louisiana and east Texas is among the highest on Earth. This is because these coasts are situated along the edge of a large sedimentary basin, the Gulf of Mexico Basin, which contains tens of thousands of feet of sedimentary strata. The weight of these strata causes the land to subside. Humans have contributed to the rate of relative sea level rive by extracting hydrocarbons and water from the subsurface.

Figure 4.4. Tide gauge records are used to estimate the rate of sea level rise over the past century. These records indicate that the relative rise in global sea level has been on the order of 15 cm (0.5 feet) in the last century. The rate has been faster in Louisiana and Texas because of higher rates of subsidence. These records also suggest that the rate of rise may have been increasing in recent decades, and this is supported by results from analysis of satellite altimetry data. Modified from Gornitz and Lebedeff 1987.

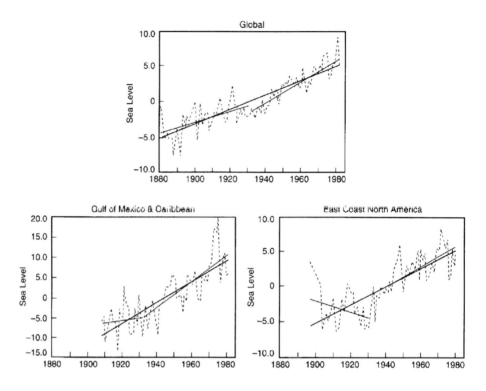

The best historical record of relative sea level change comes from tide gauges that have been recording sea level oscillations for many decades. One of the longest tide gauge records is located in Galveston Bay at Pier 21. This record shows an overall rise in sea level of about one foot over the past century. This is faster than the global average because the bay floor is subsiding. In fact, subsidence may prove to be the greatest threat to our coast if recent estimates of subsidence prove valid.

Will Galveston Island someday join the ranks of Sabine Bank and Heald Bank?

A recently published report by the National Oceanographic and Atmospheric Association (Shinkle and Dokka 2004) has significantly raised the estimates of coastal subsidence along the western Louisiana coast, which has major implications for the east Texas coast. According to this report, the rate of subsidence at the Texas–Louisiana border is about 5.4

Figure 4.5. This map shows those areas of the coast that will be submerged by relative sea level rise of three feet. Only those areas in black will remain above sea level. Geological studies have shown that coastal inundation occurs more episodically because of irregularities in the land surface. Bay and coastal environments may undergo catastrophic change as these environments are stressed by rising sea level. Thus the relative rate of sea level rise may be exaggerated, but the impact of coastal submergence may be underestimated by this figure. Also, this figure does not take into account the impact of the combined National Oceanic and Atmospheric Administration (NOAA) and IPCC worst-case scenarios. Courtesy of A. Rodriguez.

feet (approximately 1.5 meters) per century. This is an even grimmer scenario than the most extreme IPCC predictions for eustatic rise (fig. 4.3). If we take just half of the projected subsidence rate for east Texas and median estimate for global sea level rise from the IPCC results, a significant portion of the upper Texas coast will be submerged in the next century.

The reliability of the NOAA estimates has been questioned by a number of experts (*Houston Chronicle* 2005). They argue that the method used can be highly inaccurate and that the NOAA estimates are not consistent with other data, such as long-term tide gauge records for western Louisiana and east Texas, including Galveston Bay. These records indicate between 0.5 and 1.0 feet of relative sea level rise in the past century (fig. 4.4).

The NOAA report indicates the mouth of Lake Calcasieu and Sabine Lake are subsiding at a rate of 5.4 feet per century. There are beach ridges at the mouth of Sabine Lake that are 2,400 years old. Given the NOAA estimates for coastal subsidence in this area, these ridges should have subsided 120 feet since their formation, which clearly has not happened. The same can be said for Bolivar Peninsula and Galveston Island. Remember, Galveston Island has old beach deposits that date back over 5,500 years. If coastal subsidence was even in the ballpark of what is suggested by the NOAA report for western Louisiana, the Pleistocene surfaces on which

Figure 4.6. The area of rapid subsidence in Louisiana occurs mainly south of Inter-state 10 and corresponds to the area of maximum oil and gas production (shown in black). Cour-tesy of Kristy Milliken.

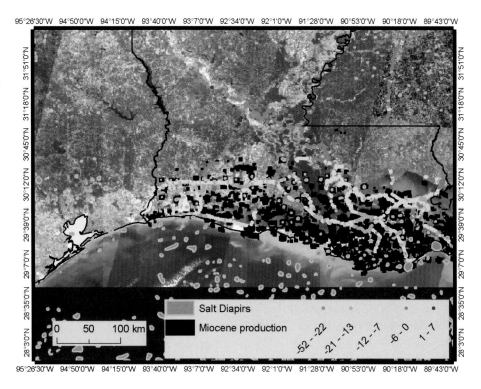

Bolivar Peninsula and Galveston Island rest would have by now subsided 200 feet or more, which has not happened. So, if the NOAA rates are valid, subsidence in the region significantly increased during the latter part of this past century. Is this possible?

One explanation for the apparent increase in subsidence this century is that it is the result of oil and gas extraction from the subsurface. A study by the Bureau of Economic Geology showed that 75 percent of known sur-face faults located between Sabine Pass and Matagorda Bay have become active in the past several decades, and this activity is correlated with the onset of nearby subsurface fluid production, both oil and water (White and Morton 1997). Figure 4.6 shows that the areas of highest subsidence in southern Louisiana correspond closely to the area of significant oil and gas production in south Louisiana.

Given the enormous potential impact of accelerated relative sea level rise on the Louisiana and Texas coast, it is not surprising that the scien-tific community is frantically assessing the NOAA estimates. If they are correct, we are in store for some major changes in our coast. Keep in mind that the NOAA study did not include east Texas. Those who are responsible for monitoring subsidence in our region have been the most outspoken against the NOAA results. One hopes they are right, but it is doubtful that the rapid subsidence stops at the Texas–Louisiana border. East Texas and western Louisiana share a lot in common as far as their geological settings are concerned. The fact is, all indications are that the rate of relative sea level rise is on the increase. Given this, we need to begin planning for the changes this increase in coastal inundation will bring about.

The Impact of Rising Sea Level on the Coast

What will be the impact of rising sea level on our coast?

One of the fundamental laws in geology is the principle of uniformitarianism, which states that the present is the key to the past. We can apply this principle to study the potential impacts of accelerated sea level rise. The rates of sea level rise in the NOAA and IPCC reports are in the range of rapid sea level rise following the last glaciation, when some rather dramatic changes were occurring along the upper Texas coast (chapter 2). That is why we study the evolution of the coast during this time interval.

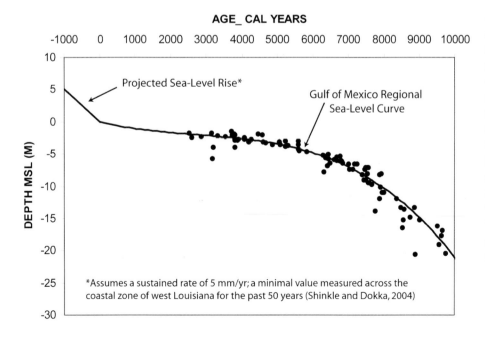

Figure 4.7. The measured rate of sea level rise for Texas for the past 10,000 years. Also shown is the projected rate of sea level rise for the next century. Note that the projected rate is roughly equal to the rate that occurred when sea level was rising most rapidly in the past, which is when our coast was experiencing the most dramatic change (see chapter 2).

It allows us to make better predictions about coastal response to accelerated sea level rise.

The net effect of rising relative sea level is landward migration of the
Gulf and bay shorelines and loss of wetlands. But the coastal response to
rising sea level in the past was far from that of simple drowning. Rather,
research has shown that the evolution of the coast was punctuated by
episodes of rapid shoreline retreat followed by intervals of slow retreat
(fig. 2.3). Likewise, the bays of the upper Texas Coast underwent dramatic
and rapid changes during their evolution (fig. 2.15). This is why it is not
possible to predict the impact of accelerated sea level rise by simply taking
topographic maps and measuring the amount of land that is inundated
by a rise of a given magnitude, such as the one shown in figure 4.5.

One impact of rising sea level on wetlands is fairly straightforward. If
the rate of subsidence exceeds the capacity of marsh vegetation to grow
upward, the wetlands will drown. Relative sea level rise is the second
leading cause of wetlands loss in Texas. Humans are the leading cause.
Chapter 2 shows how the response of bay environments to rising sea level
was punctuated by events when the system changed radically in a few
centuries. Associated with these events were dramatic reductions in the
extent of the bayhead delta and other wetlands along the bay shore. These
changes resulted from a combination of natural reductions in sediment
supply to the bays and rising sea level. Modern bayhead deltas are already
stressed by the reduction in sediment input from dam construction, so
they are already vulnerable to increased rates of sea level rise.

In some coastal areas of Louisiana, relative rise in sea level due to land
subsidence has virtually submerged thousands of acres of wetlands each
year. Texas is also experiencing wetlands loss due to land subsidence, but
at a much slower rate. This will change when sea level reaches the elevation of the low-lying wetlands that occur along the coast. This low-lying
area, the coastal plain, is the area that was submerged during the previous interglacial (120,000 BP). The low areas are coastal plain terraces
that were shaped by both rivers and waves to create their flat topography.
Much of southern Chambers County is below five feet elevation and subject to flooding during the next few centuries, or possibly by the end of this
century if the rate of sea level rise continues to increase.

The response of shorelines to rising sea level is a bit more difficult to
predict than the impact on wetlands. How fast the coast retreats land-

ward depends upon the gradient of the coast, which is rather flat in Texas and Louisiana, the material being eroded, and the rate at which new sand is delivered to the coast to fill the space created by relative sea level rise. Currently, sand supply is minimal. The combination of relatively rapid sea level rise, a gentle coastal profile, and a shortage of sand in the coastal system results in faster erosion in Texas and Louisiana than elsewhere in the country. Along our low gradient Texas and Louisiana coastlines, an annual rise in relative sea level of between 1/16 inch and 1/8 inch per year (based on tide gauge records) results in an average three feet to five feet of coastal retreat. If the rate of sea level rise doubles in the next century, the rate of coastal retreat will at least double, and this is a modest prediction. A more likely scenario is that rates will vary along the coast as they do today and as they have in the past, and that coastal retreat will be episodic (fig. 2.3).

5

Coping with Coastal Change

The road that goes nowhere.

Can beach erosion be stopped—and should we try?

Eroding Beaches or Retreating Shorelines?

Before we get started in our discussion, we need to set the record straight about "beach erosion." To a coastal geologist, the term *erosion* implies that sediment is being lost from the system. A geologist thinks about where the sediment is going and whether it is truly being removed from the coastal system or simply redistributed. Mountains erode, and much of the sediment they shed ultimately ends up in sedimentary basins. The fact is, our beaches are not eroding at a rate of several feet per year because sand is not being lost from the system anywhere near that rapidly. The shoreline is simply migrating landward. Sand that is moved offshore during storms is, for the most part, not moving seaward of the shoreface. Hence, that sand remains within the longshore transport system and ultimately finds it way back onshore. Even the sand that is being stored in back-barrier wash-over fans and tidal deltas will eventually be reincorporated back into the longshore transport system as these features are overtaken by the advancing shoreline. In fact, the rate at which sand is redistributed is fast enough that it happens in our lifetimes. For example, think about the changes that occurred off Surfside Beach when the Brazos River was diverted. Within a few decades sand from the old delta was deposited in the new delta (chapter 3). If we remove sand from an area to nourish a beach, we need to be certain that we are not taking sand that will within decades be needed to maintain the coastal profile by natural forces. Otherwise, we are creating problems that our children and grandchildren will inherit.

Where beaches have been allowed to migrate landward under the forces of nature, they have remained fairly much the same in terms of their width and profile. The Galveston State Park is a good example. Only where man has interfered with the natural migration of the shoreline have beaches significantly shrunk, such as along the Galveston Seawall. In this book I use the terms *landward migration* of the shoreline or *coastal retreat*, unless significant amounts of sediment are being removed from the system; in that case the term *erosion* is appropriate. Surfside Beach, where the sand supply of the Brazos River was diverted, is an example of an eroding beach. The beaches east of Rollover Pass are another example of eroding beaches.

Current rates of shoreline retreat along the upper Texas coast average between 3 and 5 feet per year. There are a few shorelines that are not changing significantly and other areas where they are changing much faster. The Bureau of Economic Geology (BEG) at the University of Texas is the agency responsible for monitoring rates of shoreline change in Texas (http://www.beg.uTexas.edu/coastal/intro.htm). In the past, scientists there used aerial photographs taken at different times over the past several decades to estimate rates of shoreline movement. Now they use a sophisticated method known as LIDAR (Airborne Light Detection and Ranging), which uses an airborne laser to measure very subtle changes in coastal elevation and beach location. The method is accurate to within inches, and surveys can be repeated at regular intervals.

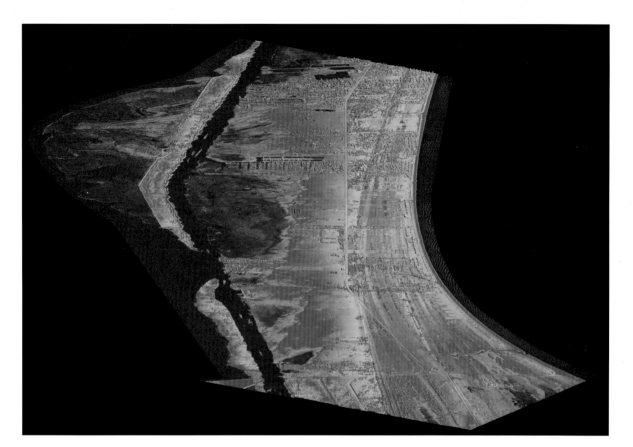

Figure 5.1. LIDAR (Airborne Light Detection and Ranging) image of Bolivar Peninsula. The tan spots are houses, and ICW is the Intracoastal Waterway. Courtesy of Dr. James Gibeaut, Bureau of Economic Geology.

Figure 5.2 shows the most recent estimates of shoreline change along the upper Texas coast between High Island and Cedar Lakes. Two areas are experiencing seaward growth, the areas on either side of the Bolivar Roads jetties and the Brazos River delta. Humans have altered both. The Bolivar Roads jetties trap sand that moves within the longshore transport system. The U.S. Army Corps of Engineers' 1929 alteration of the course of the Brazos River shifted the sand supply of the river from the Surfside areas to a new location 6 miles to the west (fig. 3.4).

Three areas of the coast are experiencing rates of landward retreat in excess of 5 feet per year. One of these is located just west of Rollover Pass at the east end of Bolivar Peninsula. The opening of the pass trapped sand that was being transported west by longshore currents. This in turn resulted in increased rates of erosion on the west side of the pass.

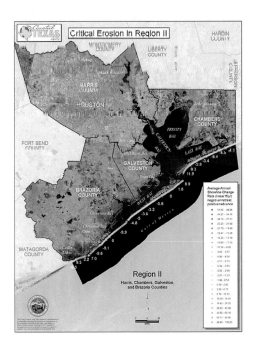

Figure 5.2. The Texas Bureau of Economic Geology is the agency responsible for monitoring shoreline change in Texas. This map shows rates of "coastal erosion" as estimated by the bureau. Positive values designate areas where the shoreline is moving seaward, and negative numbers indicate landward migration rates. The highest rates of landward migration are along the east end of Bolivar Peninsula, the east end of Follets Island, and between Surfside Beach and the mouth of the Brazos River. From Bureau of Economic Geology Web site, http://www.beg .utexas.edu/coastal/ intro.htm.

The beach just west of the Galveston Seawall is moving landward at rates of nearly 6 feet per year because little sand is making its way past the seawall and the rock groins to the east. Follets Island has a low profile and limited sand supply, and in places the shoreline is retreating landward at a rate of nearly 10 feet per year. The island also rests on water-saturated clay and is believed to be subsiding relatively fast due to compaction of these clays. The beaches at Surfside have suffered some of the fastest erosion rates of any part of the Gulf Coast since the Brazos River was diverted from this location. Since the river was diverted, extension of the Freeport jetties and deepening of the ship channel has further restricted longshore transport of sand to the area. In addition, groundwater extraction by local industries is estimated to have caused up to 2.5 feet of subsidence in the area (http://texascoastgeology.com).

What if we do nothing to slow the rate of coastal change?

There are certain facts that we must face in any attempt to alter the landward retreat of the shoreline. First, we cannot stop the shoreline from

Aerial view showing offset in the beach on either side of Rollover Pass that is attributed to increased erosion caused by sand being trapped by the pass. Photo from GlobeXplorer.

retreating landward, but we may be able to slow the process. Relative sea level will continue to rise, and sand needed to nourish our beaches is in short supply. Damming rivers has contributed to the problem, but that impact is minimal compared to the natural forces at play and to other impacts we have had on the coast. Indeed, humans have contributed to the problem, but we are beyond being able to completely rectify our impact by taking sand from areas where it is currently being trapped and moving that sand to beaches whose sand supply has been cut off. That is not to say that we should not try. Finally, the only way to slow the rate of coastal retreat is to nourish beaches, and that will require large amounts of sand and money. How much sand will be needed depends on the rate of relative sea level rise this century.

In 1959 the Texas Open Beaches Act was passed to assure all citizens the right to beach access. Texas law dictates that the boundary between public beach easement and private property is the vegetation line. The vegetation line is not always a clear-cut boundary because the vegetation line changes relative to the shoreline, especially following storms. The law is a tough one. It dictates that as the vegetation line migrates landward, so that beachfront houses are situated seaward of the line, those houses must be removed at the owner's expense. While this seems harsh, especially to beachfront property owners, the law exists because lawmakers recognized that the shoreline does continuously retreat landward. Without such a law, all beaches would eventually become private property littered with collapsing houses, and public beach access would no longer exist. For those who plan to build on the beach or buy a house on the beach, a good rule of thumb is to build on land that will still be theirs for the time they own their home. For stretches of West Galveston Island and Follets Island, that is not very long.

The General Land Office (GLO) is the agency responsible for moni-

Cement pillars are all that remain of a condominium that was under construction when Hurricane Alicia struck in 1983. At that time, the new structure was located approximately 200 feet landward of the shoreline. The rate of shoreline migration is nearly 10 feet per year at this location. Even if the condos had survived the storm, the beachfront units would now be located seaward of the vegetation line.

The west end of Galveston Island is experiencing rates of coastal advance of between 3 to 5 feet per year on average. Here, beachfront houses are about to be overtaken by the advancing vegetation line.

toring changes in the vegetation line and determining which houses to condemn. That is not a pleasant task, and the GLO has received a lot of criticism for their actions. Property owners argue that the erosion is a temporary problem that will reverse itself with time. As discussed earlier, that is not going to happen. Some political leaders have adopted the same attitude, which has not helped the situation. One politician even said that the government should not enforce the law, meaning the Open Beaches Act, until the cause of the erosion problem has been determined and the process reversed. That statement was made nearly ten years ago. Meanwhile, the beach has continued to march landward. These statements reflect the general lack of understanding of our coast. This is not acceptable for politicians who will help decide its destiny. Landward retreat of the shoreline is not a recent phenomenon. There is absolutely no scientific basis for arguing that the rate of coastal retreat will decrease. Indeed, as discussed earlier, the rate will increase if the rate of sea level rise increases. And finally, it will take an act of God, not politicians, to reverse the process. The question is not whether natural phenomena will slow the rate of coastal retreat but what, if anything, can be done to delay the inevitable?

Even with the most optimistic prediction, which is that the rate of coastal retreat will equal that of the past few decades, the shoreline on Galveston Island will retreat landward an average of 150 feet by the year

Figure 5.3. This photo-graph shows the loca-tions of the shoreline in 1956 and the projected 2056 line.

1956 Shoreline

2056 Shoreline

2056. If these predictions seem unrealistic, stand at the west end of the Galveston Seawall and look toward the west. When the seawall was completed, the beach at this location extended nearly 100 feet seaward of the wall. In the 1960s, at low tide one could drive on the beach along the seawall. The shoreline has since retreated to the base of the seawall. But, to the west, where there is no protective barrier, the beach retreated land-

Figure 5.4. This aerial view of the west end of the Galveston Seawall provides another perspective on the erosion that has occurred west of the wall. At the current rate of coastal retreat, the highway will again be overtaken by the advancing shoreline by the end of this century, perhaps sooner if major storms occur during that time. The key point here is that while the beach offshore of the wall was destroyed, there is still a beach west of the seawall. There the beach was not eroded; it simply moved landward.

ward an additional 400 feet, or a total of 500 feet. Clearly the seawall has done its job at preventing the highway and adjacent property along the east end of the island from being destroyed. The price has been the loss of the beach. The only beaches that occur along the seawall are those that were nourished and those that are at the far east end of the island where the jetties have blocked the longshore transport of sand.

Construction of the seawall is not the only case where humans have tried to stop the retreat of the shoreline at the expense of losing the beach. In the past, property owners have attempted to protect their property by constructing bulkheads between their houses and the retreating shoreline. As the shoreline retreated, no beach remained between the bulkheads and the Gulf. On either side of these features, the width of the beach remained constant with time. Only where the structure was placed has the beach disappeared. This is one of the reasons why the Texas Open

View looking west from the west end of the Galveston Seawall. In the 1950s one could drive onto the beach from this part of the highway.

Beaches Act exists. Lawmakers who passed this law understood that placing structures at the water's edge would ultimately result in a loss of public access to beaches.

One way to get around the concerns about placement of hard structures, such as bulkheads, on the beach has been to use sand socks, fiber tubes (geotextile tubes) filled with sand. The logic here must be that if you can press down on it, it is not a hard structure. While they may not be "hard" structures in the legal sense, common sense tells us that there is no difference between sand socks and bulkheads. The sand socks have the same effect in that they provide a temporary barrier to the landward migrating shoreline and vegetation line. But, like bulkheads and seawalls,

The construction of bulkheads to protect private property is now discouraged by law. This photograph shows why. Here, the bulkhead has served its purpose of protecting beachfront property, but the beach has continued to march landward. Over time, beach access has been eliminated.

Bulkheads cannot stop the shoreline from retreating landward. Eventually, the beach is eliminated.

Figure 5.5. View looking east at the western end of the Pirates Cove sand socks. The beach seaward of the sand socks has decreased in width by as much as 30 feet, which is consistent with measured rates of shoreline change for this area (see fig. 5.2).

they do not slow the rate of retreat and may even contribute to the problem. So, sand socks serve only one purpose, protecting beachfront property. That is why the Federal Emergency Management Agency (FEMA) helps to underwrite their cost. Meanwhile, the shoreline continues to retreat landward, and the beach grows narrower and steeper. As a result, the energy of storm waves is focused on a smaller portion of the beach with a steeper profile, making it more subject to erosion. If maintained, the areas where sand socks exist will have no beaches, just as the west end of the Galveston Seawall has no beach.

No matter how they are shaped, sand socks don't grow new sand grains, although some of the people who promote sand socks seem to imply that when they argue that beaches actually grow in front of the socks. In fact, if we were to place sand socks from the west end of the seawall to San Luis Pass, in just a few decades there would be no beach, just waves splashing against the socks. And what about the natural coastal habitats that are being destroyed? Allowing the beach to shrink between the

Sand socks, like the ones shown here, are fiber tubes filled with sand. They are currently being used to "slow the rate of coastal erosion," as their proponents say. Unlike bulkheads, they are not discouraged by law, even though they have the same Impact on the coast, which is to cause the beach to shrink.

swash zone and sand socks is slowly removing these habitats, which in-cludes shorebird habitats and turtle nesting areas. Turtles just can't seem to get a break in Texas.

One way of justifying sand socks has been to argue that they are arti-ficial dunes. But attempts to maintain natural dune vegetation and habi-tats on sand socks have failed, while natural dunes have slowly vanished, having no place to migrate landward. This is why the Texas General Land Office has gone on record with the statement that geotextile tubes are not dunes and do not protect public beaches. The GLO has also recom-mended that the socks remain covered with sand and vegetation, which has proven to be quite difficult. The GLO has made other recommenda-

Scenes like this are becoming more common along the coast and provide a lesson for those who believe that sand socks stop the landward migration of the shoreline.

Easy come, easy go. During November 2004, Pirates Beach sand socks were covered with sand from a nearby sandpit. Within three weeks, much of that sand had been washed away.

tions that make good sense, including that the sponsor of the project is responsible for maintenance and removal of the tubes once they begin to restrict access. The GLO further recommends that the sponsor must replace or enhance public access. We shall see if their recommendations are followed.

This photograph was taken less than one month after the previous photo (p. 81) was taken. This is all that remains of the sand that was placed on the beachside of the socks. The culvert in the foreground was added for runoff, which is blocked by the sand socks.

Once the decision to install sand socks is made, the long haul of maintaining them begins. This is a battle that Mother Nature continues to win.

Wetlands Loss

Wetlands are our most threatened and valuable coastal resources. An estimated 98 percent of all commercial fish and shellfish are dependent upon wetlands in either their life cycle or as part of their food chain (http://www.nmfs.noaa.gov). This does not take into account the role wetlands

An added problem with sand socks is that they prevent proper offshore drainage, so culverts must be installed. This concentrates offshore drainage, so that gullies form and roads are undercut.

play in the life cycles of waterfowl and other creatures, their importance for human recreation, and their role in protecting the bay shore from erosion during storms. It was indeed alarming that during the post–World War II era, the rate of wetlands loss in this country was on the order of 300,000 acres per year. In Texas alone, nearly half of our wetlands had been destroyed, either naturally or by human intervention, by the mid-1980s. Those of us who remember days when bay fishing was better can attest to the impact of wetlands loss.

The trend toward total destruction of our wetlands was curbed when the Clean Water Act was passed and when state agencies began to regulate wetlands use and change. The Protection of Wetlands Executive Order

View looking north across pristine wetlands on the east end of Follets Island.

No. 11990, passed in 1977, directs all federal agencies to avoid, to the extent possible, long- and short-term damage to wetlands and to avoid support of projects that would adversely impact wetlands. If this law had not been passed and the destruction of wetlands curbed, the resulting impact would have been unthinkable. Unfortunately, the laws are shortsighted; they protect existing wetlands but do little to protect future wetlands. To add to the problem, we are now struggling with natural loss of our remaining wetlands, a battle that we are currently losing. The Coastal Wetlands Planning, Protection, and Restoration Act (CWPPRA) was passed in 1990 to allow federal agencies to take a more proactive role in wetlands restoration. To date, Louisiana, which currently loses about 8,760 acres of wetlands per year, has had the most aggressive wetlands restoration program. Texas is only beginning to undertake wetlands restoration projects, and the road ahead is a long one if we are going to regain even a fraction of the wetlands lost to human and natural causes. Currently, the United States Geological Survey (USGS) is assessing the impact on wetlands in the Galveston Bay area, particularly as a result of subsidence.

In addition to monitoring rates of "coastal erosion," the Texas Bureau of Economic Geology has also monitored rates of erosion along our bays' shorelines. Since our bay shores are mostly inhabited by wetlands, these rates reflect rates of wetlands loss—and those rates are astonishing. Figure 5.6 shows an example of measured erosion rates for West Bay, on the

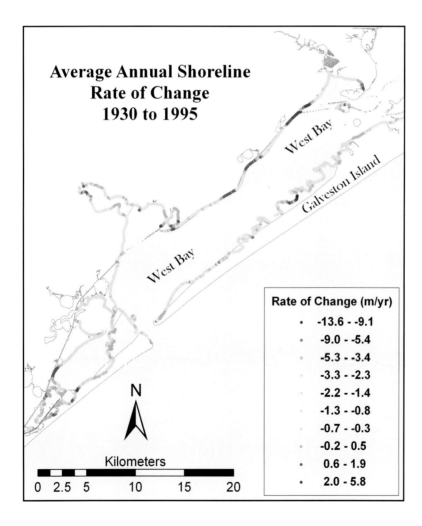

Figure 5.6. Shoreline erosion rates for West Bay. From the Bureau of Economic Geology, http://www.beg.utexas. edu/coastal/intro.htm.

landward side of Galveston Island. On average, the bay shoreline is eroding at rates in excess of 5 feet per year, and locally rates are as high as 40 feet per year. Similar rates of erosion occur in East Bay and in Galveston Bay.

Why are wetlands disappearing at such high rates?

Wetlands like those shown here are vanishing at an alarming rate, in part because humans have eliminated wetlands' ability to migrate landward under the constant pressure of rising sea level.

Like the beaches of the upper Texas coast, wetlands are feeling the effects of relative sea level rise. Humans are contributing to the problem. Wetlands have had to contend with relative sea level rise for thousands of years, but our past actions have now put them at a major disadvantage. For example, the natural response of wetlands to the rising sea is to migrate to higher ground. But in many areas, such as along the south shore

Sand socks have served some beneficial purpose. On the south shoreline of West Bay, sand socks are used to protect the marsh from erosion. The greatest erosion takes place when waves generated by strong winds from the north slowly eat away at the subsiding wetlands. Like beach erosion, this process was occurring long before development on the island began, but people have built on those areas where the wetlands would have migrated.

of West Bay and East Bay and along virtually the entire western shoreline of Galveston Bay, humans have built on those areas where wetlands could migrate. Other impacts are less obvious. For example, for wetlands to sustain themselves they need a certain amount of sediment to help build the framework on which they exist. Any action that reduces the input of sediment to wetlands will reduce their ability to exist. Damming rivers has decreased sediment supply to wetlands, and construction of highways and the Intracoastal Waterway, for example, has reduced natural sediment pathways to the wetlands. As far as wetlands are concerned, hurricanes are not necessarily a bad thing. A hurricane every now and then helps keep the supply of sand moving into the back barrier, and these wash-over deposits become wetlands. Coastal development is preventing this process from happening.

If something is not done to reduce the rate of wetlands loss, the impact will have far-reaching effects. That is why the National Marine Fisheries

Aerial view of wetlands restoration project at the Galveston State Park.

Service, in conjunction with the Texas General Land Office, the Galveston Bay Foundation, and many volunteers, launched a campaign to protect and even restore wetlands. It will be a long, hard battle, but one that we must fight if we are going to save our most valued and threatened coastal asset. It is too bad that we spend so much money and effort trying to protect beachfront property and so little trying to protect our wetlands.

The U.S. Army Corps of Engineers has played a leading roll in sustaining wetlands by placing dredged material from navigation channels on the wetlands through the Regional Sediment Management Program. However, funding for this program has been reduced.

One of the most effective ways of protecting wetlands is to build a barrier, using sand socks or rock breakwaters, around them to protect against

These circular piles of sand will become new wetlands once grass is planted.

Sand socks are used to protect a restored marsh from waves. This example is from the recently completed wetlands project near Jamaica Beach.

New development on the west end of Galveston Island is allowing little room for wetlands to migrate landward as rising sea level slowly drowns them.

wave erosion. In some cases sediment is pumped onto the wetlands to create new areas for growth and expansion. Volunteers plant grasses, and new wetlands are born. This is not an easy job, so these volunteers deserve a word of thanks.

Despite our efforts to reestablish wetlands, they are still being threatened by developers. The problem is that current laws only protect existing wetlands. They do not allow for the natural migration of these wetlands as relative sea level rises and forces them landward. A case in point is the far west end of Galveston Island, where development is currently taking place right to the edge of the existing wetlands. Given current rates of shoreline migration in this area (fig. 5.6) these wetlands will be gone in about a decade.

One of several cattail marshes on Galveston Island. These important freshwater habitats are rapidly disappearing along the upper Texas coast.

While we are on the topic of wetlands, let us not forget the freshwater wetlands (interdunc swales) that are so essential for the wildlife along our coast. On Galveston Island, these wetlands are slowly being filled, or the vegetation is removed so that houses can be built on or near them.

The Vanishing Dunes

In addition to adding beauty to the coast, dunes are nature's defense line against storms, although they are not much of a match for powerful storms. Hurricane Alicia destroyed much of the natural dune line on

The vanishing dunes. View facing south of natural dunes and dune vegetation near Surfside.

Galveston and Follets Island. The dunes will reestablish themselves after a storm, but usually in a more landward position. If there are houses landward of the dune line, the dunes will have no place to go. This is why natural dunes are becoming more and more of a rarity along the upper Texas coast. Like beaches and wetlands, they need room to migrate. As we lose our dunes, we are also losing a unique biological habitat with a diverse plant and animal assemblage, including some endangered species.

A number of local, state, and federal agencies are involved in the protection of dunes.

This photo shows why the dunes are disappearing with time. They have no place to migrate landward as the shoreline retreats in that direction.

There are laws to prevent people from removing sand from dunes and destroying the vegetation on the dunes, but they are not always enforced.

The Beach Access and Dune Protection Laws are intended to protect the coastal dunes of Texas. As with the wetlands situation, laws that are intended to protect dunes are shortsighted because they do not allow for the natural landward migration of dunes as the shoreline retreats landward.

In an effort to remedy the rapid loss of natural dunes, artificial dunes have been constructed using a variety of methods, none of which have worked. One example of a failed attempt is the "straw dune" project at Pirates' Beach. Bales of hay were piled on the beach and covered with sand. There was also an attempt to vegetate the dunes, which include a sprinkler system. The straw and sprinkler system ended up on the beach after the dunes were destroyed by a tropical storm. Later efforts to construct artificial dunes have seen little more success. These more recent efforts

Dune construction project at Jamaica Beach in June of 2006.

Hay bale dunes at Pirates Beach. The lesson learned here is that straw is no defense against storm waves.

Geotubes don't make good dunes.

have relied on sand socks as the foundation for the dunes. The result has been limited periods when vegetation is able to grow on the dune followed by removal of the sand and vegetation during storms. Once the commitment is made to construct artificial dunes, the long and costly process of maintaining these dunes begins.

The Impact of Hurricanes on the Coast

On average, a major hurricane strikes the upper Texas coast every six years. From http:// www.nnvl.noaa.gov/ hurseas2005/katrina 1615z-050829-1Kg12.jpg.

What impact would a Category 5 hurricane have on the upper Texas coast?

This book so far has focused on those changes in our coast that occur over relatively long time periods. What about the catastrophic impact of tropical storms?

A Category 5 hurricane is one with sustained wind speeds of greater than 155 miles per hour. Such a storm could bring a storm surge of 20 feet, possibly as much as 25 feet in some areas, to the upper Texas coast. Coastal flooding would occur as far away as 150 miles on either side of the eye of the storm, and heavy rains, likely more than 12 inches within a day, would cause widespread flooding well inland of the coast. Saltwater wash-over into wetlands would considerably damage wetlands. The 2005 landfall of hurricanes Katrina and Rita taught us all a lesson about the damage that powerful storms can have on the coast.

History of Hurricane Impact on the Upper Texas Coast

Historical records describing the damage inflicted on the upper Texas coast by major tropical storms and hurricanes date back to 1527 (See http://www.srh.noaa.gov/lch/research/txhur.php). Since 1837, when the storm record becomes more complete, twenty-nine major storms have struck the upper Texas coast. That is an average of one major storm every six years.

The first storm to inflict heavy damage on Galveston Island was in 1837. That storm destroyed nearly half the houses on the island, along with the original Tremont Hotel. In 1821 the town of Valasco, which was located at the mouth of the Brazos River near what is now Surfside Beach, suffered heavy damage from a storm that destroyed nearly half the town. Valasco's population was more than 25,000 at the time, but there is no accurate account of how many lives were lost as a result of the storm.

By far the best-known storm to strike the upper Texas coast was the Great Storm of 1900. That storm virtually destroyed the city of Galveston and killed most of its inhabitants. It is estimated that as many as 6,000 lives were lost. Following the 1900 storm the seawall was constructed, and the elevation of the downtown area was raised approximately 5 feet. The sand needed to raise the island and construct the seawall was taken from a large sand pit at the center of the island, which is now occupied by Offats Bayou. Since then the island has experienced other storms, but the seawall and elevation changes have helped spare the city of Galveston from major damage and loss of lives.

Since 1959 eight named storms (hurricanes) have struck the upper Texas coast, inflicting heavy damage. These include Debra (1959), Carla (1961), Cindy (1963), Edith (1971), Claudette (1979), Alicia (1983), Bonnie (1986), and Rita (2005). Hurricane Carla in 1961 was a powerful storm that caused considerable damage to the upper Texas coast, even though the eye passed far to the west near Port Lavaca. In the Galveston Bay region, the storm tide from Carla ranged from 9 to 15 feet, and low-lying coasts flooded as far away as High Island. During Carla, Bolivar beaches

The city of Galveston after the 1900 storm. Courtesy of the Rosenberg Library.

After the 1900 storm, the people of Galveston rebuilt the city, which included raising the elevation of the island five feet and constructing the Galveston Seawall. Courtesy of Rosenberg Library.

suffered up to 60 feet of erosion. Had Hurricane Carla passed over Galveston Island, its entire west end would have been beneath 10 to 15 feet of water.

The most devastating storm to strike the Galveston area in recent years was Hurricane Alicia, which made landfall in August 1983. The path of

Offats Bayou occupies what was a large sandpit where sand was excavated to raise the elevation of Galveston Island and construct the seawall following the Great Storm of 1900. From the U.S. Geological Survey.

the storm was through Galveston Island, and the destruction to private property was beyond any natural event since the 1900 storm. The storm demolished many buildings between High Island and Surfside. Most damage was to beachfront homes. Alicia destroyed most of the remaining dunes on the west end of Galveston Island and caused tens of feet of beach erosion (Dupre 1985). Most of the sand that was removed from the beach during the storm returned in a few months.

After wreaking havoc on the coast, Hurricane Alicia traveled up Galveston Bay, where it continued to inflict damage. Afterward pleasure boats were piled high all along the bay shore, and even downtown Houston suffered damage. My own sailboat was left perched on the highest point in Seabrooke, Texas.

On September 24, 2005, Hurricane Rita made landfall at the Texas and Louisiana border, inflicting heavy damage in a swath that extended from Galveston Island to New Orleans. For more than twenty-four hours, computer models projected Rita's landfall at Galveston Island. This led to evacuation of the island, Bolivar Peninsula, and outlying coastal towns, the largest mass evacuation in U.S. history. It will also go on record as

This aerial photograph, taken just days after Hurricane Allelu made landfall on the upper Texas coast, shows severe damage to houses along the coast. Note that the most severe damage was inflicted on houses located nearest the beach. This is because of undercutting of houses near the beach as the beach profile was lowered.

the most chaotic evacuation, because major highways were clogged with those trying to evacuate the Galveston and Houston areas. Those who suffered the heaviest damage in places like Port Arthur and Beaumont will look upon Rita as the worst storm in history. But the impact could have been even greater. At its peak, Rita was packing wind speeds of 175 miles per hour, but those had decreased to 120 miles per hour when the storm made landfall. Had Rita's intensity not diminished, the consequences would have been unthinkable.

How Storms Impact Coasts

None of the storms that have struck the upper Texas coast during the past century have been especially powerful (Category 4 or 5) storms—a frightening thought. We have never experienced a Category 5 hurricane, nor do we want to. We have had some close calls. For example, Hurricane Gilbert in 1988, a Category 5 storm, appeared as though it would make landfall on the upper Texas coast, though ultimately it came ashore on the northeast coast of Mexico. More recently, powerful Hurricane Katrina struck New Orleans and the Mississippi and Alabama coasts.

The initial impact of a major hurricane is flooding caused by water being forced landward by strong winds. This is called *storm surge,* and coupled with astronomical tides (storm tide), water levels could reach 30 feet above normal. The storm tide during Hurricane Katrina was 26 feet. The

Holly Beach, Louisiana, before and after Hurricane Rita made landfall. From the U.S. Geological Survey.

June 16, 2001

September 28, 2005

≋USGS

storm tide allows large waves to travel inland, leveling trees and houses in their path. As the storm moves inland, the second attack on the coast begins as water that was forced inland by the storm flows offshore and undermines buildings.

The combination of high storm tides and strong winds can move sig-

Tropical Storm Frances, a relatively small tropical storm, caused considerable erosion and damage along the upper Texas coast.

nificant quantities of sand from the beach. Even relatively small tropical storms, such as Dean, Josephine, and Frances, have resulted in as much as 60 feet of shoreline retreat in a single event (http://texascoastgeology. com.). Tropical Storm Frances produced storm tides that were 3 to 5 feet above normal and 10- to 15-foot waves that pounded the coast for two and a half days, resulting in considerable beach erosion. This demonstrated that a relatively small storm that moves slowly across the coast can cause considerable damage.

One can learn a lot about the potential impact of a major storm by viewing the damage inflicted on areas that have been less fortunate. During the summer of 2004, Hurricane Ivan, barely a magnitude 4 storm, struck the west Florida and Alabama coasts with such devastating impact that the area looked as if it had been bombed. Even massive cement structures were leveled, undermined as the waves and tide lowered the beach

Pirates Beach, shortly after Tropical Storm Frances struck the upper Texas coast in 1998. The greatest damage from the storm was undercutting of structures by combined waves and storm tides.

profile. They stand as a reminder that even the strongest structures can fall prey to the undercutting of the beach during storms.

When Hurricane Katrina struck the Louisiana, Mississippi, and Alabama coasts, the resulting devastation was a wake-up call for everyone, including those who supposedly plan for such emergencies. One of Katrina's impacts on the natural environment was to destroy roughly half the Chandeleur Islands, which, fortunately, are undeveloped. The combined impacts of both Katrina and Rita destroyed most of the homes on the west end of Dauphin Island, Alabama. It is unlikely that the island will ever recover. Dauphin Island is smaller and narrower than Galveston Island, but could Galveston Island withstand a similar double blow?

Coastal geologists have long known that the process by which coasts migrate landward involves creation of new equilibrium profiles (see chapter 1). It is also well established that new profiles are created during storms. The historical record of this process is limited to only a few centuries, but the geological (rock) record contains many examples of landward migration of coasts by way of this process. When humans tamper with the system by placing breakwaters and sand socks on the beach, the result is a beach profile that becomes too steep with time. Large storms establish a new equilibrium profile, and in the process, undermine any

One impact of storms is undercutting of houses as the storm surge relaxes and the water that was piled in back-barrier areas rushes back offshore.

structure that stands in the way. With time, even the Galveston Seawall will fall to this process.

Another impact of hurricanes is to breach barrier islands and peninsulas. More often than not, the areas that are breached fill with sand soon after the storm, but sand that is washed across the barrier through storm channels is lost from the beach.

Geological Evidence of Past Storms

There is geological evidence that very powerful storms have struck our coast on more than one occasion in prehistoric time. The evidence exists in the form of breaches in the barriers (storm surge channels) and wash-over fans of sand moved from barriers into back-barrier bays during storms. In general, the wider the barrier, the less likely it will be breached. Although there are storm deposits in the back-barrier bays, to date no detailed work has been aimed at reconstructing storm history from these deposits.

Beach ridges on the seaward side of Bolivar Peninsula are younger than 800 years old, and they rest on a prominent erosion surface. These observations indicate that at some time prior to 800 years ago, Bolivar Peninsula was virtually submerged by a major hurricane. It took centuries to grow back to its pre-storm size. The Karankawa Indians were

*These before and after
photographs show
the damage caused by
Hurricane Ivan, which
struck the west Florida
and Alabama coasts in
2004. The condominium
on the left was destroyed
by undercutting of the
beach while the adjacent
house, which stands on
pilings, suffered rela-
tively minor damage.
From the U.S. Geological
Survey Web site, http://
coastal.er.usgs.gov/
hurricanes/ivan.*

When Hurricane Ivan struck the west Florida and Alabama coasts, it breached the barriers, providing an example of this process of barrier evolution. Aerial photographs taken before, top, and after, bottom, show a breach in the barrier. From the U.S. Geological Survey Web site, http://coastal.er.usgs.gov/hurricanes/ivan.

Storm wash-over channels and fans on the back side of Galveston Island record powerful storms that were strong enough to breach the island when it was younger and narrower.

This wash-over fan on the bay side of Bolivar Peninsula is the product of powerful storms that breached the barrier and delivered large amounts of sand into the back-barrier bay. From the U.S. Geological Survey.

Figure 6.1. This aerial view of the west end of Galveston Island shows the city of Jamaica Beach near the center. Beach ridges can be traced in the landscape for about a mile west of Jamaica Beach, where they are no longer visible. West of this point, the island is narrow and shows evidence of having been breached by storms within the past 1,000 years. Photo from GlobeXplorer.

living on Galveston Island and Bolivar Peninsula when these storms struck, and it is unlikely that any of them survived.

There is also evidence, in the form of storm surge channels and wash-over deposits, that Galveston Island was once breached by hurricanes. But beach ridges that extend virtually uninterrupted along the island to just west of Jamaica Beach characterize the current landscape on the island. The far west end of the island lacks prominent beach ridges but is dissected by many storm surge channels. Since we know the age of beach ridges that have not been breached by storms, we know how long it has been since the eastern half of the barrier was breached. The youngest beach ridge that has not been breached is about 2,400 years old. The far west end of the island has been breached on several occasions in the past 1,000 years.

Predicting Storm Impact

With any major storm the first concern is human safety. As for that, we must trust we are in good hands, but hurricanes Katrina and Rita suggest we should not rely too much on others to watch out for our safety. What about damage to the coast? How much damage will be done, and where will the greatest impact occur? Armed with observations of hurricane impact on the Gulf Coast in historical time and with geological observations of past storm impacts, we are able to make predictions about the impact of a Category 5 hurricane on the upper Texas coast.

Sabine Pass would suffer major damage from a Category 5 hurricane, and the storm surge would extend as far inland as Port Arthur and its industrial complex. Hurricane Rita taught us this. The beaches west of Sabine Pass are sand starved, and there are no barrier islands. Hence, the shoreline would simply step landward, probably several tens of feet. And don't be fooled by the name High Island. If I lived on High Island and a hurricane were headed my way, I would be headed north well in advance of the storm, because the roads north would be flooded with only a few feet of tidal elevation.

The weak link in the shoreline west of High Island is at Rollover Pass, a man-made cut through the narrowest part of the peninsula at its eastern end. The pass provides an outlet for moderate storm surge and post–storm flow, but in a Category 5 hurricane, the inlet could be widened.

My greatest concern about the impact of a Category 5 hurricane is its effect on the industrialized, low-lying areas in the upper reaches of Galveston Bay. When a hurricane moves inland over a bay, the tendency is for the storm surge to increase as the onshore flow of water is concentrated within the bay. The damage done to the I-10 bridge across Pensacola Bay during Hurricane Ivan and across Lake Pontchartrain during Hurricane Katrina are reminders of the damage that a strong storm surge can do well inland from the coast. Computer models indicate that, in a major storm, the upper part of Galveston Bay, which includes the Port of Houston and all the industry that surrounds it, would be submerged by a storm tide that could easily exceed 20 feet. Also impacted by flooding would be cities like Texas City, Kemah, League City, and the Clear Lake area, including the Johnson Space Center. In a recent article in the *Houston Chronicle*, officials estimated that as many as 600,000 homes could be flooded.

There is no question that Galveston Island will suffer major damage if it is struck by a Category 5 hurricane. The first impact would be flooding of the island by a storm tide that would likely exceed 15 feet. This plus 155-mile-per-hour winds would destroy most beachfront homes. It is not unreasonable to expect that the shoreline along the west end of the

Aerial view of Rollover Pass shortly after Hurricane Alicia.

island would advance landward at least 50 feet and that the island might be breached in several locations at the far west end. The San Luis Pass Bridge would likely suffer major damage, and the storm surge would inflict heavy damage on the industries in the Chocolate Bayou area.

Follets Island would be the hardest hit by a powerful hurricane because the island is narrow and low, and because it consists of only a few feet of sand resting on clay. The main effect of a major storm would be significant over-wash and landward migration of the island and associated expansion of the wetlands, similar to what occurred on Dauphin Island, Alabama, following Hurricane Katrina. The island would likely breach in a number of locations, and it would probably take many years for the island's meager longshore sediment supply to fill the breached areas. County Road 3005 (Bluewater Highway) would be destroyed along much of its extent.

Another area of significant storm impact would be the industrialized Freeport and Lake Jackson areas. The Freeport area is "protected" by levees, but Hurricane Katrina taught us not to rely on those levees to ward

Figure 6.2. This computer-generated model depicts the extent of flooding in the low-lying industrial area in the upper part of Galveston Bay, including the Houston Ship Channel and Johnson Space Center (JSC). From http: www. spaceref.com/news/viewsr.html?pid=8144.

Photograph taken along Highway 90 in Gulfport, Mississippi, after Hurricane Katrina inflicted heavy damage to the coast in the summer of 2005. This photograph was taken in June of 2006, nearly a year after the storm.

off storm surge. Hurricanes have breached barrier islands, so man-made levees are not likely to survive their impact. When this happens, the threat of chemical spills will be high.

Does a hurricane have any positive impacts?

While hurricanes can cause significant damage to coasts, they may play a beneficial role as well. During storms, sand washes across barrier islands into back-barrier marshes. The process is more prevalent where the barriers are narrow, such as along Follets Island. The sand that washes across the barriers provides plants with a framework on which to grow and keep pace with the relative rise in sea level. The wetlands on the bay side of Galveston Island and Bolivar Peninsula are shrinking in size because they are being drowned by subsidence, and sand wash-over is virtually nonexistent. It is possible that back-barrier wetlands would actually increase in area in the aftermath of a major hurricane as storm wash-overs supply more sand. However, the initial impact would be a loss of wetlands that are buried by the wash-over and destruction of vegetation by waves and exposure to more saline water during storm surge.

Figure 6.3. The extent of flooding that would occur in the Texas City and Galveston area in the event of a Category 5 hurricane with a 22-foot storm tide. Galveston Island would be completely submerged except for the taller buildings in the city. From http://www .spaceref.com/news/ viewsr.html?pid=18144.

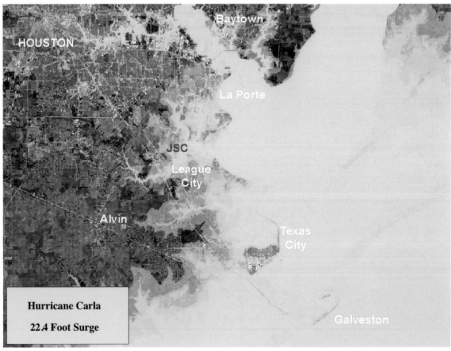

A breach in Follets Island that nearly cut through County Road 3005 during Tropical Storm Frances.

The Freeport industrial complex. Virtually all of the area shown in this photograph would be submerged by the storm surge of a major hurricane. From the U.S. Geological Survey.

Aerial view of the Freeport channel in 2005. The petrochemical complex shown in this photograph would sustain heavy damage if struck by a major hurricane.

Is it true that hurricanes will increase in number, size, and intensity as global warming increases the temperature of Gulf waters?

While tropical storms are less severe than hurricanes, they occur more often along the upper Texas coast and have had greater cumulative impact than hurricanes during the past century. The damage inflicted by hurricanes Frances, Ivan, Charley, Dennis, Wilma, Katrina, and Rita in 2004 and 2005 left the Gulf Coast from Bolivar Peninsula to the Florida Panhandle in virtual ruins, although Tropical Storm Allison, which struck in 2001, still ranks as the most costly storm to strike the upper Texas Coast. These were unprecedented years in terms of hurricane dam-

Sand that washes to the back side of the barriers during storms helps the wetlands by providing a framework on which marsh grasses can grow and keep up with subsidence that would otherwise drown the marsh.

age to the Gulf Coast, although 1900 still ranks as the worst year in terms of loss of human lives.

Within days after Hurricane Rita, a flurry of newspaper and magazine articles heralded the combined Katrina and Rita storms as unique events that might have been caused by global warming. The fact is, the scientific community is still divided as to whether global warming has had any influence on hurricane frequency, although many agree that the warming trend in the atmosphere and sea surface temperatures has fueled larger and more dangerous storms; indeed this has long been predicted by many global climate models. A recent article in *National Geographic* magazine argued that the United States could be facing "decades of coast-crushing hurricanes because of changes in ocean circulation patterns and an increase in sea-surface temperatures" (http://news.nationalgeographic. com/news/2001/07/0719_hurricanes.html). This article may have been inspired by an earlier article in *Science,* "The Recent Increase in Atlantic Hurricane Activity: Causes and Implications" (Goldenberg et al. 2001). The authors of the *Science* article show that the number of major hurricanes affecting the Caribbean Sea and the Atlantic has increased dramatically since 1995. The number of hurricanes affecting the Caribbean has increased fivefold, and on a global scale, the number of Category 4 and 5 storms has increased relative to Category 1, 2, and 3 storms in recent

Figure 6.4. Sea surface temperatures in the Gulf of Mexico are measured from space, which provides much-needed regional coverage. The image on the right is the mean summer temperature, and the smaller inset image shows winter temperatures. The curve at the top shows the 20-year record of surface temperatures since this type of data has been collected. While temperatures have increased very slightly over this time period, this increase is not generally thought to be a significant factor in regulating hurricane strength. From http:// poet.jpl.nasa.gov.

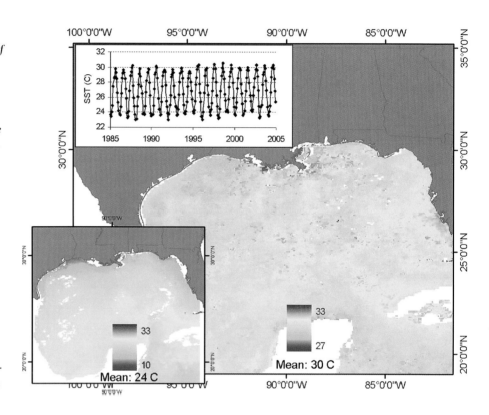

years. All indications are that the trend will continue for the next 10 to 40 years. Goldenberg and his colleagues attribute this pattern to natural climate cycles, but the scientific community continues to debate the role global warming will play in enhancing this cycle. What is important for those of us who live along the Texas coast is that the Gulf of Mexico is expected to experience only minor differences in hurricane frequency and magnitude, although it is hard to convince oneself of this given the number of impacts along the Gulf Coast in 2004 and 2005. One must remember, however, that hurricanes are born in the Atlantic and Caribbean, and the number of storms formed during any given year does not depend that heavily on sea surface temperatures. Once hurricanes form and move into the Gulf, they feed on heat from surface waters, so the magnitude of storms will increase if surface water temperatures increase and as greater moisture is added to the atmosphere. The record of sea surface temperatures in the Gulf of Mexico for the past 20 years shows little evidence of a significant increase. Thus 2004 and 2005 may have been anomalous years. We can only hope that this is the case.

Combating Coastal Change

Houses constructed too close to the shore, like sand castles, have a short life span.

What can be done to slow the landward advance of the shoreline?

In reality, there is only one way to slow the rate of coastal retreat, and that is to place more sand on the beach. This is referred to as *beach nourishment*. Trying to engineer our way around the problem has had a detrimental impact, namely the loss of beaches.

A number of my colleagues in coastal geology would argue that we should do nothing to slow the process of coastal retreat. I think that there are areas where this advice would be well taken. However, we have to accept the fact that the citizens of Galveston are not going to allow their city to be washed out to sea without a fight. What if a Category 5 hurricane were to strike the island and destroy the seawall? Would we rebuild a bigger and stronger seawall? A major storm would also move sand into back-barrier settings and offshore where that sand may be lost from the longshore transport system indefinitely. How much coastal real estate would we be willing to relinquish? I am enough of a realist to know that if the worst happens, there is likely to be a frantic search for sand. Had Hurricane Rita maintained its strength and course, Galveston Island would have suffered major damage, and we would be hard pressed to replace sand that would be lost from the island. That is not to say that we have not spent a lot of tax dollars looking for sand resources. We just have not gotten much for our investment.

Searching for Sand Resources

First we have to accept the fact that sand resources for beach nourishment are in short supply along the upper Texas coast. I have been asked many times, "So why not pump sand from offshore onto the beach? It works in Florida, so why not in Texas?" The difference is that the Florida shelf is covered by sand, and our shelf is covered by mud. We have the Mississippi River to thank for this.

Along the upper Texas coast, beach quality sand is mostly confined to the shallow part of the upper shoreface, generally in less than 15 feet of water and within a half mile from shore (fig. 1.2). It makes little sense to dredge sand from the shoreface to nourish a beach because that removes sand from the longshore transport system, which nourishes the beach naturally. If we take sand from the longshore transport system and place it on the beach at one location, we are removing sand that would end up on the beach in another location. Thus the effect is the same as building jetties. Furthermore, removing sand from the shoreface only steepens its profile, and nature will work hard to reestablish that profile by reclaiming that sand.

What about taking sand from the lower shoreface? Sand that occurs in the lower part of the shoreface is finer than beach sand. This very fine sand is carried in suspension by waves and currents more easily than the sand that forms our beaches. That is why the very fine sand resides farther offshore, where it is not constantly influenced by waves and wave-generated currents. If we place very fine sand on the beach it will simply

The beaches that now exist along the seawall have been artificially nourished by sand pumped from offshore. This is a costly exercise, but the biggest problem with sustaining our beaches by this means is a shortage of offshore sand resources. The photo at the top was taken shortly after the beach was nourished in 1995 with the pipeline still in place, and the one at the bottom was taken in 2005.

be transported right back out to the lower shoreface, probably in the first tropical storm. Furthermore, the very fine sands of the lower shoreface are typically interbedded with mud (see chapter 1). So the quality of sediment that could be acquired from the lower shoreface is not likely to meet the standards of beach-quality sand.

We Texans do enjoy our beaches. Each summer tourists flock to the beaches of Galveston to enjoy the water and sun, bringing much needed revenue to the island. This beach was nourished in 1995. The beach has been narrowing each year, but those who enjoy it would say the money was well spent.

How much sand will be needed to stabilize our beaches?

Before we get too far along with trying to stop "beach erosion," we need to think about how much sand is needed and what areas need to be nourished. This is not an easy question to answer, but we must make some effort to answer it if we are going to continue pouring tax dollars into temporary solutions, such as sand socks, with the argument that more permanent solutions are in sight. To determine how much sand is needed, we first need to decide what we will consider a stable beach and how much, if any, shoreline change we are willing to accept.

Is there such a thing as a stable beach?

Most coastal geologists would likely respond that a stable beach is one whose profile and adjacent shoreface is in equilibrium with wave energy. When the beach erodes, its profile is lowered and it becomes steeper (see also chapter 1, fig. 1.6). Before the beach can be returned to its stable equilibrium profile, its elevation must be raised and the offshore profile rees-tablished. Otherwise, sand that is pumped onto the beach will ultimately be transported back offshore to become part of the shoreface.

To date, there has been only one large-scale beach nourishment proj-ect on the upper Texas coast. In 1995 the city of Galveston financed a nourishment project for an approximately 4-mile stretch of beach east of Sixty-first Street. An estimated 535,500 cubic meters of sand was pumped directly onto the beach using a pipeline dredge. This project cost $7.65 per cubic meter of sand plus $1 million dollars for mobilization and demobili-zation. Much of that sand is now gone from the beach. Most city officials would say that the project was a success. It certainly has kept people com-ing to the island to enjoy the beach. The problem is, sand that is needed to nourish the beach is limited.

The sand resource that was used to nourish the beach in 1995 was, in fact, taken from the shoreface off East Beach. But in this case the shoreface has been artificially nourished with sand trapped behind the South Jetty. That sand resource was virtually depleted by the 1995 nourishment proj-ect. There are other sand resources in the Bolivar Roads area, between the jetties, and to the east of the North Jetty, and these will likely be used in the near future. Other sand resources exist offshore, but they are too far to be pumped directly onshore. This means that the sand will have to be transported to the nourishment area using a hopper dredge, and the cost of this type of operation is much higher than that for pumping sand directly onto the beach from more proximal borrow sites using a pipeline dredge. A hopper dredge is a ship that is capable of withstanding offshore wave conditions, storing sand that is collected from the sand source, and transporting that sand to the areas where it is needed for beach nourish-ment. The cost of this type of nourishment project is estimated to be four times the cost of the 1995 nourishment project. And the term "borrow site" is telling: nature is going to take back what we take; it is just a matter of time.

In summary, to nourish our beaches in a way that will last for a decade or more will require a lot of sand. Beach nourishment is like everything else: you get what you pay for—and sand is not cheap. The only way to do the job right is to restore the original profile of the beach. Simply pumping sand onto the beach from immediately offshore is not a solution.

Sand Resources

How do we find sand resources for beach nourishment,
and what progress has been made so far?

The fastest way to identify potential offshore sand bodies is to use seismic surveys to map prominent features such as old river channels. However, seismic data will not show whether these features actually contain sand. Sediment cores are needed to verify the occurrence of sand deposits and to test the quality of the sand for beach nourishment. Both methods require a lot of time and money.

One advantage we have in Texas is that we have a lot of oil company platform borings on the shelf. These are drill sites that are collected for engineering purposes wherever a new offshore platform is about to be located. Unfortunately the sites are usually not cored, but detailed descriptions of the drill cuttings are made, and these data have proven useful in the absence of core. There are hundreds of these platform boring locations on the shelf, and they have proven to be a valuable resource for locating offshore sand bodies, especially when combined with seismic data.

The first sand resource study along the upper Texas coast was conducted by the U.S. Geological Survey in the late 1970s and involved mostly seismic reconnaissance work between San Luis Pass and Bolivar Roads (Williams et al. 1979). Later, various projects were conducted as collaborative efforts between the Minerals Management Services, the federal agency responsible for monitoring and regulating offshore mineral

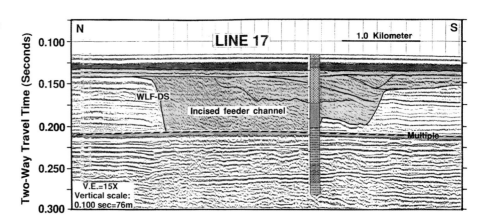

Figure 7.1. Example of a seismic profile from the continental shelf that crossed an old river valley. A platform boring from the valley reveals that it is largely filled with sand.

Seismic tow fish, which emits sound source for seismic imaging of the seabed, being lowered into the water.

The Rice University research vessel Lone Star *taking a sediment core off Galveston Island. The* Lone Star *was used to acquire several thousand miles of seismic data and hundreds of sediment cores off the Texas coast.*

resource exploitation, and the Texas Bureau of Economic Geology (BEG). Unfortunately little new data were acquired, except regarding the areas of Sabine Bank and Heald Bank. The results of these efforts were summarized in two reports issued by the BEG in 1993 and 1995.

I have no idea how much of our tax dollars have been spent in the name of offshore sand resource assessment, and I probably do not want to know. What I do know is that it has not always been spent wisely. In fact, improving the quality of the work that is done may be the biggest obstacle to eventually nourishing our beaches.

A prime example of a poor investment of our tax dollars is a recent study conducted under the watchful eye of the U.S. Army Corps of Engineers. In their recently released and much anticipated report on sand resources of the upper Texas coast, the corps identifies several potential sand resources (Finkl, Andrews, and Benedet 2004). The key word here is potential because the sand bodies identified in this report are still not proven to exist. Unfortunately, the methods used in this study simply were not suited for the task. The approach used involved divers pushing a tube with a jet of water into the sediment and observing the material washed out of the hole. After the tube was pushed as far as possible, the divers would return to the boat to describe what they had observed. Imagine trying to see through the murky water, let alone having the mental capacity to record the washings in the order they were expelled from the hole. I have tried jet probing on land where one can see the washings, and the result was pretty dismal when we calibrated our observations with actual core samples from the same location. The corps study did not attempt to calibrate with actual core samples taken at the same location. Even if the jet probe method gave a rough approximation of sand thickness below the sea floor, there is no way to measure the amount of sand versus mud or the grain size of the sand. To add to the problem, some of the sand resources identified in this study were within the shoreface, and we have already discussed the problems with this.

As one might guess, the Corps intends to spend more money in the name of sand resource studies. If you don't get it right the first time, spend more money. This would be amusing if our tax dollars were not being used to fund this work. When it comes to sand resource studies, there is little or no accountability for poor performance. Also, we spend a lot of money to find out what we already know. In this chapter I will try to summarize what we already know about sand resources so that hopefully we will not continue to reinvent the wheel. We don't have the money or time to reeducate those who are supposed to be solving our problems.

I first got started in offshore work off the coast of east Texas in 1991, when Rice University purchased a research vessel, the *Lone Star,* to conduct marine geological research in the Gulf of Mexico. The *Lone Star* studies were never intended to search for sand resources. The work was funded by a consortium of oil companies who were interested in using the more recent strata of the Gulf of Mexico as analogs for oil and gas exploration in the deep subsurface. The American Chemical Society's Petroleum Research Fund also helped fund our research. During the 1990s my students and I collected more than 15,000 miles of seismic data on the Texas and western Louisiana continental shelf, steaming along at 5 knots for more days than I care to count. The *Lone Star* also had an awesome coring capability—a 15-foot long tower that could be lowered to the seafloor, take a core, and be back on deck within a half hour. Our captain, Mark Herring, who had a knack for making things work in difficult conditions, designed the rig. One of my graduate students named our coring system the "Tower of Power." We collected hundreds of cores with the Tower of Power.

The Rice University drilling vessel, R/V Trinity.

Figure 7.2. The Rice University Coastal Research Group Web site (http://gulf.rice. edu/coastal) provides descriptions of sediment cores collected over the past twenty-five years. The map on the left is the index map that appears on the main page of the Web site. Each of the areas outlined in blue can be viewed in detail on the Web site by clicking on the boxes. The red dots represent core locations. Core logs are viewable by clicking on the red dots. This is the most extensive data set for the upper Texas coast and includes hundreds of sediment cores. Additional information, including seismic profiles and sand body locations on the shelf, can be found at http:// gulf.rice.edu.

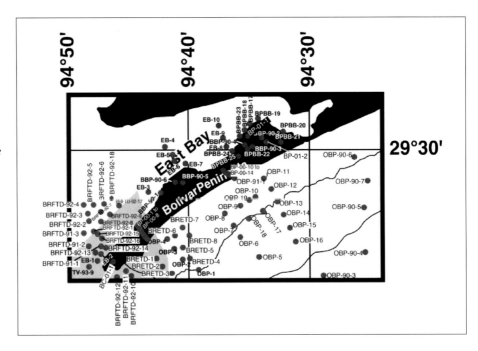

In 2000 we retired the *Lone Star* and acquired a small drilling barge, the *R/V Trinity*. We used the *Trinity* to acquire seismic data and long cores in shallow coastal waters. The National Science Foundation funded this research, which was aimed at predicting the coastal response to climate change and sea level rise in the next few centuries (see chapter 3). The Texas General Land Office also supported our research. The data acquired from these vessels is the most extensive data set ever collected for offshore sand assessment in Texas. The entire data set is available on the web (http://gulf.rice.edu/coastal). The data are available in ArcIMS at http:// tx.coast.beg.utexas.edu/website/ricecore/viewer.htm.

Near Shore Sand Resources

Are there proven offshore sand resources for beach nourishment?

A few sand bodies exist near the coast that could be used to nourish beaches in the immediate future or to help repair damage from a major storm. It would be wise to conserve at least some of this sand for storm

Aerial photograph of Rollover Pass showing sand that has accumulated on the bay side of the inlet. From the U.S. Geological Survey.

damage abatement. Here are these resources as they occur from east to west along the coast:

Rollover Pass

Rollover Pass, an artificially maintained inlet between the Gulf of Mexico and the far eastern end of East Bay, was opened in 1956. Since that time, sand moving within the longshore transport system has flowed into the pass and accumulated in a prominent bar on the bay side of the pass. An estimated 10 million cubic yards of sand exists in the bar and adjacent spoil islands (http://texascoastgeology.com). However, recent work has shown that the sand is mixed with mud, so the total volume of beach quality sand may be less than these earlier estimates indicated (Moya 2006). There is justification for using this sand to nourish beaches in the area, since the sand was diverted from the natural longshore transport system. Also, removal of the sand may have minimal impact on wetlands,

Aerial photograph showing sand accumulations in Big Reef, which occurs along the east end of Galveston Island on the north side of the South Jetty. Photo from GlobeXplorer.

although this must be evaluated. The way to prevent loss of sand from the longshore transport system is to close Rollover Pass.

Bolivar Roads Area

Before the Bolivar Roads jetties were built, most of the sand that is now trapped by the North Jetty was deposited in the Bolivar Roads tidal delta. Some sand made its way to the west to nourish Galveston beaches. An estimated 28 million cubic yards of sand has been trapped on the beaches east of the North Jetty (fig. 1.4), (http://www.texascoastgeology.com). Additional sand is located inside the jetties and offshore.

After the Bolivar Roads jetties were constructed, the bar was significantly modified by changes in the tidal flow through the Bolivar Roads tidal inlet (fig. 1.5). In addition, contaminated dredge spoil from the ship channels was dumped on the ebb tidal delta, covering portions of it. The net effect of building the jetties has been to disrupt the natural longshore

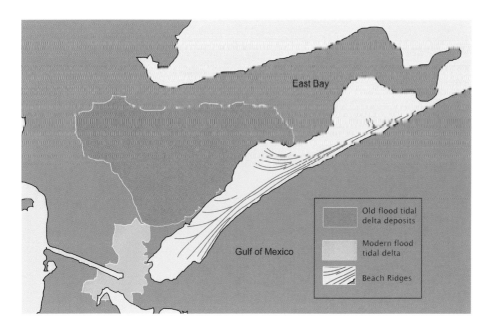

Figure 7.3. The pre-modern Bolivar Roads flood tidal delta may be a viable source for sand resources, but to date only a few sediment cores have penetrated deep enough to sample the delta. These cores did sample sands that are of good quality for beach nourishment.

transport of sand, trapping sand in the shadows of the jetties. On the positive side, that sand now occurs where it can be more easily acquired for beach nourishment. In the past, much of this sand would have ended up in the flood tidal delta.

Centuries before the jetties were constructed, the flood tidal delta was much larger. The inlet and delta complex migrated toward the west as Bolivar Peninsula grew in that direction. The delta was also larger due to the greater tidal flow that occurred before the inlet narrowed. Figure 7.3 shows a reconstruction of the former tidal delta, based mainly on seismic data. A few long cores have actually sampled the old tidal delta, and these have sampled good quality sand (fig. 2.13). That sand is now buried beneath several feet of bay mud (fig. 2.15). This may be the best large sand resource near Bolivar Peninsula and Galveston Island. The modern tidal delta contains more mud than the old delta did, but good quality sand does occur in the area just east of Pelican Island.

San Luis Pass Area

The San Luis tidal delta is the largest natural tidal delta in Texas. It is what engineers refer to as an "unimproved" tidal inlet, a term that has never ceased to amuse me. The delta contains most of the sand that has

View facing south across the San Luis Pass tidal delta. Note the large sandbars on both the bay side and Gulf side of the bridge.

Aerial view of the San Luis Pass ebb tidal delta showing sandbars and tidal channels. Photo from GlobeXplorer.

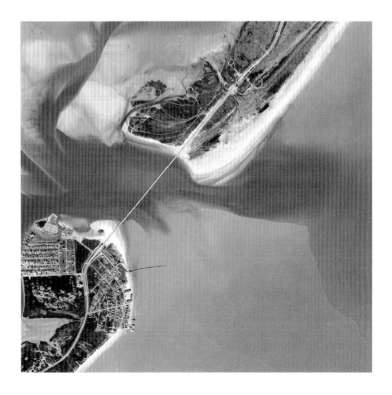

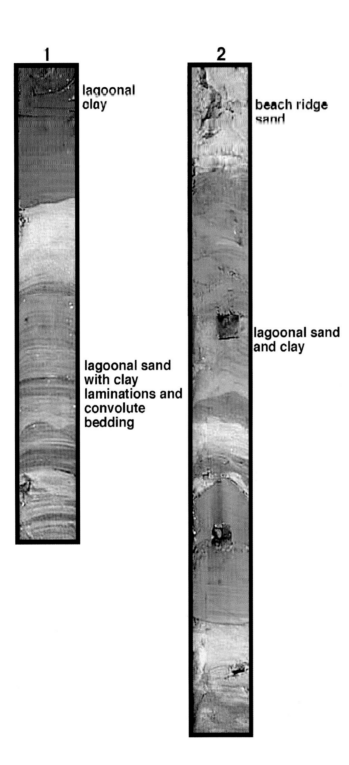

1

lagoonal
clay

lagoonal sand
with clay
laminations and
convolute
bedding

2

beach ridge
sand

lagoonal sand
and clay

Figure 7.4. Sediment cores from the onshore part of the Brazos Delta have sampled sand interbedded with clay, reflecting the various environments that have existed within the delta during its relatively short life span. This, plus the fact that wetlands cover much of the delta plain, make it an unsuitable sand resource for beach nourishment.

been eroded from Galveston Island by natural processes. Most of this sand occurs in the flood tidal delta, but part of the tidal delta is a protected wetlands and bird sanctuary. Removal of just the sand from bars within the flood tidal delta would potentially alter the flow of water into and out of West Bay and could have an adverse impact on the bay. It would certainly influence future wetlands development. The flood tidal delta should be considered off-limits for sand exploitation.

The sizable ebb tidal delta at San Luis Pass is considered by some a viable sand resource for beach nourishment. However, removal of sand from the ebb tidal delta could also alter the natural tidal circulation between the Gulf and West Bay. The impact could be quite significant. Furthermore, the sand within the ebb tidal delta will ultimately make its way west, where it is needed to maintain Follets Island.

Brazos River Delta

The modern Brazos Delta is another potential source of sand for beach nourishment. In fact, one could argue that removing sand from the delta to nourish the beaches at Surfside Beach, where the delta was originally located, is justified. But, before we get too carried away with this idea, it is important to recognize that the delta is not comprised entirely of sand. In recent years my team has collected many sediment cores from the delta, both onshore and offshore, so we have a good idea of where sand exists within the delta. Cores from the onshore part of the delta sampled sand interbedded with clay. So the onshore delta is not a good sand resource. Regardless, it is now a wetlands and is therefore protected.

The cores collected in the offshore portion of the Brazos Delta sampled mostly clay, except for the sandbar that occurs just offshore of the river mouth. The sand that comprises the bar is of good quality for beach nourishment. But much of this sand will ultimately be transported onshore and be accreted to the mainland to form new wetlands or will be transported to the west where it will help nourish beaches. This is one of the few areas of the upper Texas coast where wetlands are growing without human intervention.

Offshore Sand Resources

The Rice University data set has been used to map the major sand-prone features on the Texas continental shelf (see http://gulf.rice.edu). This work has revealed a number of potential sand resources. Some of the

The modern Brazos Delta is located 6 miles from Surfside. This is one of the few places where wetlands are expanding. Taking sand from the delta would reduce this wetlands growth.

sand bodies, which include old river valleys, deltas, sand banks, and tidal deltas, are large enough to nourish beaches of the upper Texas coast for centuries, but the cost of exploiting these would be great.

Old River Valleys

During times when sea level was lower, rivers extended their valleys across the shelf. These valleys contain sand that may someday prove to be a viable sand resource for beach nourishment. The major river channels have been mapped on the Texas continental shelf, so much of the work has been done, although more detailed work is needed to map smaller channels. But not every channel is filled with sand. In fact most Texas rivers deliver more silt and clay to the coast than they do sand. The trick is to determine which valleys contain suitable sand for beach nourishment. We can predict this to some degree if we apply a bit of science to our prospecting, something that is not always done, I have discovered. If channels are confined to the coast and continental shelf—that is, if they do not extend far inland—they likely contain little or no sand because there

Holly Beach, located west of Cameron, Louisiana, is the site of a beach nourishment project, undertaken to preserve wetlands, that used sand from old river channels to nourish the beach. Offshore breakwaters were constructed to help preserve the beach. The sand used to nourish the beach was a better quality sand than the original beach sand. Since this photograph was taken, Hurricane Rita struck the area and washed most of this sand into the wetlands.

is no source for that sand. Channels cut during the lowstand in sea level have the greatest potential for containing beach quality sand, because when sea level was low, rivers had the steepest gradients and therefore the greatest capacity for transporting coarse sand.

Rice graduate students Mark Thomas and Kristy Milliken mapped in detail the ancestral Calcasieu, Sabine, and Trinity river valleys on the continental shelf. These valleys have also been sampled in a number of locations by oil company platform borings and are known to contain sand. That portion of the ancestral Trinity River valley that is located within 50 miles of the coast contains an average thickness of 25 feet of good quality sand. The ancestral Sabine River valley, which merges with the Trinity River valley about 15 miles offshore, contains slightly less sand in its lower portion than the Trinity valley. Again, the problem is that this sand is covered by an average of 80 feet of mud.

The volumes of sand within the Trinity and Sabine valleys are measured in billions of cubic yards. That is enough sand to nourish our beaches for a long time. The other good news is that the quality of the sand is quite high, based on the minimal sampling that has been done. If this sand were used to nourish our beaches, they would look like the Miracle Strip of Florida. The challenge is how to mine these sands from beneath thick layers of mud.

During the last lowstand of sea level, the Brazos and Colorado rivers extended their valleys across the continental shelf and nourished large deltas on the outer shelf. These old river valleys were mapped by Dr. Ken

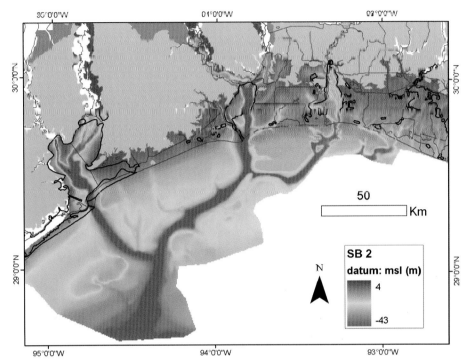

Figure 7.5. The potential sand resources for the upper Texas coast include old river channels and terraces, deltas, tidal deposits, and sand banks. This map shows the locations of the ancestral Trinity, Sabine, and Calcasieu river valleys that were formed when sea level was lower and the shelf was exposed. Courtesy of Kristy Milliken.

Abdulah as part of his PhD research at Rice. Platform borings indicate that the Brazos valley contains beach quality sand, but it is buried beneath mud of variable thickness.

The Colorado River formed two offshore valleys during the last lowstand of sea level (fig. 7.7). Of all the large river valleys on the shelf, these hold the greatest promise as sand resources. Seismic data and borings indicate that the valleys are virtually filled with sand. Analyses of the sand indicate that it is of excellent quality for beach nourishment.

River Deltas

There are a number of old river deltas on the Texas shelf, and they contain large volumes of sand (fig. 2.1). But most of these sandy deltas are located on the outer shelf, more than 50 miles from shore, so exploiting them would be costly. The only exception may be the former Colorado River delta that is located offshore of Caney Creek and the Colorado River mouth. Jennifer Snow studied the delta for her master's thesis research at Rice. Using seismic data and sediment cores collected with the *Lone Star*,

Figure 7.6. Old river channels contain sand that could be used for beach nourishment, but the sand is not always located near the seafloor where it is easily exploited. This seismic section across the ancestral Sabine River valley was collected offshore of Bolivar Peninsula. A long core through the middle of the valley shows that the sand is confined to the lower part of the valley, beneath 90 feet of mud, so the sand is below the reach of conventional dredges. Unfortunately, this is the case for most of the old fluvial channels that occur offshore of the upper Texas coast.

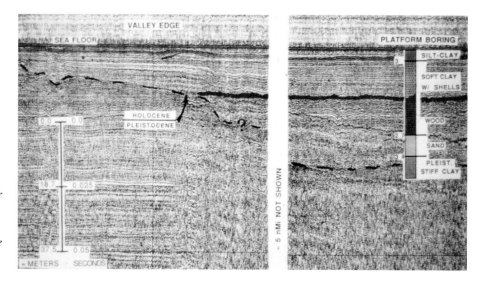

Figure 7.7. This map shows the locations of the ancestral Brazos and Colorado river valleys on the continental shelf. The rivers cut these valleys as sea level fell to the edge of the continental shelf during the last glaciation. The map is based on a detailed study of these old rivers and their associated deltas by Rice PhD student Ken Abdulah.

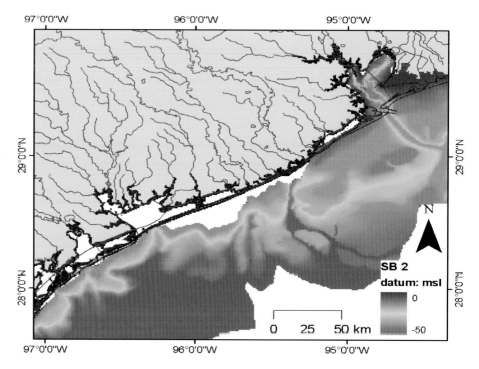

Snow constructed the map shown in figure 7.8, a thickness map of the delta.

There is enough sand in the old Colorado delta to nourish the beaches of the upper Texas coast for decades. These sands are rather coarse compared to modern beach sands, and if they were used to nourish our beaches, the result would be beautiful white sand beaches.

Sand Banks

Two large sand banks, Sabine Bank and Heald Bank, exist on the inner part of the continental shelf offshore of the upper Texas Coast. There are two smaller banks, Shepard Bank and Thomas Bank, but to date these have not been sampled. These banks are the remains of former barrier islands that were drowned in place by the advancing sea (see chapter 2). Detailed studies of the Sabine and Heald banks by the Bureau of Economic Geology and by Rice University have shown that they contain significant volumes of sand. The BEG estimates that the two banks contain 1.8 billion cubic meters of sand.

Tidal Delta Deposits

As sea level was rising during the past several thousand years, the Trinity River valley was flooded to create ancestral Galveston Bay, which migrated to the north within the valley as sea level rose. At times during the overall flooding of the valley, the rate of flooding and landward migration slowed long enough for barriers to be established along the flanks of the valley. Sand banks are remnants of former barriers. Where there are barriers there are inlets, and where there are inlets there are sandy tidal deltas. Such was the case for the old Trinity Valley shorelines, and the locations of the old inlets and associated tidal deltas have been established by careful mapping of the valley (fig. 7.9). This work was done by Dr. Mark Thomas for his PhD dissertation, and the results have withstood more recent examination by another Rice student, Dr. Tony Rodriguez. Figure 7.10 shows a cross section down the axis of the valley and illustrates how these old tidal deposits are identified. Note the different seismic characteristics of the tidal delta deposits (Seismic Section 1), which shows large-scale cross bedding, which results from the shifting of tidal channels, and the more horizontally layered bay deposits (Seismic Section 2). Platform borings indicate that the tidal delta deposits are sand-prone, but cores are needed to determine exactly how much sand exists and its grain

Figure 7.8. The Colorado River delta contains enough high quality sand to nourish our beaches for many decades. This map shows the thickness (in feet) of the delta. Borings through the delta have sampled sand, but this sand is interbedded with mud, so more detailed study will be required before this sand resource could be used. Courtesy of Jennifer Snow.

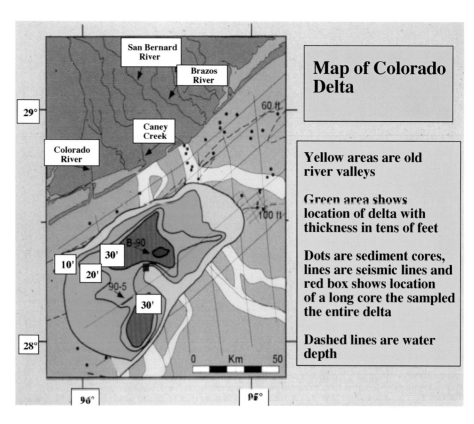

Sand from a drill core sample from the ancestral Colorado River valley. The sand is of excellent quality for beach nourishment.

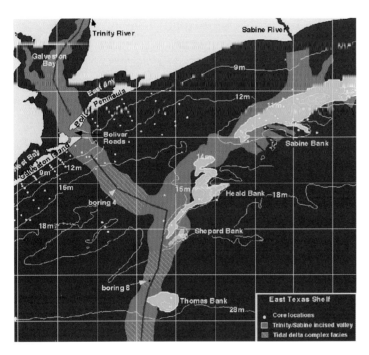

Figure 7.9. This map of the upper Texas continental shelf shows the location of sand banks and the ancestral Trinity and Sabine river valleys. Also shown are the locations of old tidal deltas that occur at shallower depths in the Trinity valley. The red line marks the location of an axial profile of the valley, part of which is shown in fig. 7.10.

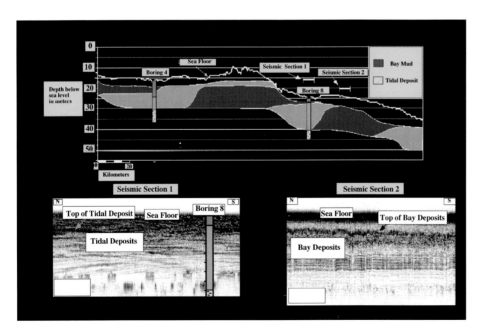

Figure 7.10. This profile along the axis of the offshore Trinity River valley was constructed from seismic data and sediment cores and illustrates how sandy tidal deposits are identified from these types of data. Note that the tidal deltas are characterized by large cross beds (seismic section 1) in contrast to muddy bay sediments that are characterized by horizontal layering. Platform borings 4 and 8 verify that the tidal deposits are sandy, although core was not recovered at these sites. These old tidal deltas may someday prove to be valuable sand resources for beach nourishment.

size. What we do know is that these sandy deposits occur much nearer the seafloor than do the sandy river deposits that occur in the base of the valley. The locations of the tidal deposits are shown in figure 7.10. When the day comes that major sand resources are needed to sustain Galveston Island, these old tidal deltas will surely warrant further investigation.

Onshore Sand Resources

Several years ago we began considering the fact that old river channels may extend beneath the coast and could be viable sand resources for beach nourishment (see chapter 2, fig. 2.11). The Sabine and Trinity river valleys, which lie beneath Sabine Lake and Galveston Bay respectively, have already been discussed. But what about the Brazos and Colorado river valleys? If we could determine the locations of these old river valleys, then we could look for extensions of these valleys just offshore of the beaches that need nourishing. The idea was that even if the sand within these valleys is buried in mud, some of that mud may have been removed by transgressive ravinement (see chapter 1). Whether a valley extends offshore or not will depend on when the valley was occupied by the river and where the coast was at that time. Only those valleys that were active during sea level lowstands, when the shoreline was located offshore, would extend any significant distance offshore. We focused our attention on the Brazos valley, since the Brazos is the largest river and the one closest to where sand is needed.

Old River Valleys

In an ongoing PhD project at Rice University, Patrick Taha has collected literally thousands of water well descriptions from the onshore portions of the ancestral Brazos River valley. These descriptions are made every time a water well is drilled, and they include information about the depths at which sand was encountered. Taha has also acquired seismic data and ground-penetrating radar profiles to try to isolate the valley location. Taha's work showed that the Brazos River has occupied several valleys since the lowstand in sea level (fig. 2.11), but only one of these valleys, the Big Slough Valley, is old enough to extend seaward of the present shoreface. That valley is located beneath the modern San Luis ebb tidal delta. Additional work is needed to pinpoint the exact location of the valley and to determine if it contains beach quality sand.

Taha's work focused on the valley occupied by the Brazos River during

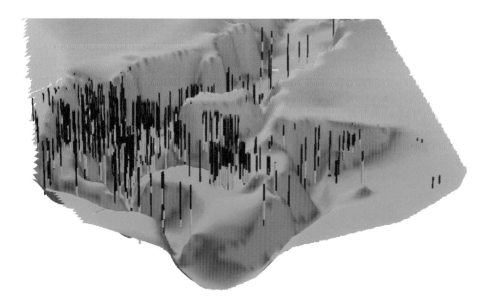

Figure 7.11. This figure shows a digital elevation map of the incised Brazos river valley. Each of the red and yellow lines is a water well that sampled the valley fill. The red indicates clay, and the yellow is sand. Courtesy of Patrick Taha.

the lowstand in sea level. His results show that this valley extends beneath the current Brazos River and that sand is confined to the lower part of the valley, typically beneath 40 to 75 feet of red clay. The sand within the upper part of the valley is interbedded with clay, but there is beach quality sand in the lower part of the valley. Our seismic records from offshore show the valley on the inner shelf, but we have not acquired any cores from the offshore portion of the valley.

Sand Pits

Until 2004 the strategy used to maintain artificial dunes and sand socks was to use sand from sandpits on Bolivar and Galveston islands. Some have suggested that we pull even more sand from these sandpits to nourish beaches and maintain sand socks. The argument is that this would be a temporary solution necessary only until offshore sand resources are tapped. The reality is that digging pits and trucking sand to the beach is analogous to throwing buckets of water on a forest fire—only sand is much more scarce than water on the island. It is a pretty good rule of thumb that if you see a dump truck driving down the coast, you are probably watching your tax dollars being dumped into the Gulf.

In November 2004 Pirates Beach property owners trucked large amounts of sand from a borrow area on the island to cover sand socks and

Drilling a core behind Follets Island to locate the old Brazos River channel that extends beneath the island.

regrade the beach. Within two weeks most of the sand had been washed away (see chapter 5). That sand was immediately moved into the long-shore transport system. Some will likely move back onshore during summer months, but property owners to the west will be the main benefactors. At this rate, it will take a lot of sandpits to maintain the existing sand socks on the island.

In 2004 the city of Galveston passed an ordinance that stopped the mining of sand on the island, much to the dismay of some property owners. But city leaders are to be commended for doing the right thing. If one were to measure the sand needed to stop erosion, or even slow the process, the sandpits would stretch from one end of the island to the other.

Why has the current system failed, and how might we fix the problem?

During the past several years I have attended annual meetings convened by the Texas General Land Office that are intended to convey the latest information about how to preserve the coast and where sand resources for

Water-filled sandpit near Pirates Beach. This is one of several sandpit operations on Galveston Island and Bolivar Peninsula.

nourishing our beaches exist. In the past these meetings took on a bit of a carnival atmosphere, with engineering companies exhibiting their latest products to stop beach erosion and claiming to have all the answers to our problems. At the September 2005 meeting, the GLO decided to focus more on what has been and is being done in the way of coastal research in Texas. I attend these meetings to learn how our tax dollars are being spent in the name of coastal preservation. The truth is, we have spent a lot of money and made little progress toward identifying viable sand resources to nourish our beaches. We spend a lot of money putting sand on isolated beaches in the hopes that it will remain there a few years. We fund people to tell us what we already know. We fund surveys that are poorly planned and that generate little in the way of real results.

When it comes to getting a bang for our buck, we simply are not getting much return on our investment. The current approach is to fix leaks with the small amount of funds that are available. We need to replace the plumbing. Putting sand on a one-mile stretch of beach or sand socks in front of a row of houses does not solve the problem. We must think on a more regional scale. But this will not be possible with the present, poorly coordinated approach. There are too many government agencies doing their own thing and not enough communication between these agencies.

The search for sand involves a fair amount of science, and creative engineering. I have spent my career in the academic world, where good science and engineering depend on the peer review system. When I write a proposal to a funding agency to do research, that proposal goes out for review, meaning that it is sent to colleagues who are experts in the field. They evaluate my past record for generating results from National Science Foundation (NSF)-funded projects, the validity of the scientific approach, and the likelihood of achieving solid results. If I am lucky enough to be funded—and most scientific proposals are not—I go about doing the project with the knowledge that I am accountable. When I complete the research, I am expected to present my results at national and international meetings before experts in the field and to publish my results in peer-reviewed journals. If I stand before a body of respected scientists and present bad results, the immediate penalty is professional embarrassment. If I submit those results for publication, they will be rejected, usually in not so kind a fashion. Publish or perish is not a bad thing—it is academia's natural selection process. The ultimate penalty for doing bad science is that one will no longer get grants, and someone else will get a shot at the research funds.

Why shouldn't there be a similar system of selection and accountability to insure that our tax dollars are spent wisely and, more importantly, that we are making progress either toward identifying the sand resources that are needed or at protecting existing beaches and wetlands? If we make the agencies who fund the work accountable, they will make the people who do the work accountable. There lies the problem. The state of Texas needs a better review system for proposed projects and more stringent accountability. Until that happens, we will likely continue to see a poor return on our investment in coastal preservation.

What is needed in Texas is an independent body that will keep track of the different government agencies involved in protecting and monitoring our coast to avoid duplication of effort. That same body should screen proposals and assign a team of reputable coastal geologists, engineers, and economists to review proposals and evaluate projects to assure that taxpayers get a fair return on their investment.

The state of Texas has just learned that it will receive about $250 million, which could be designated for coastal work. Just how much of that money will be dedicated to beach nourishment is yet to be decided, but it is a sure bet that at least some of it will. If that money is used to increase the size of the bucket brigade so that more buckets of water are thrown on the forest fire, there will be little to show for it. Instead the money should be used to pick an offshore sand resource and begin nourishing beaches at a regional scale. We simply have to get away from the current practice of every man for himself and begin working as a team to solve our coastal erosion problems.

Development of the Coast

The construction of the Galveston Seawall was a bold statement that we humans intended to stop the shoreline from moving landward. We lost that battle, but the struggle continues.

Who is watching after our coast?

On paper, our coast is in safe hands. The Texas Open Beaches Act guarantees beach access for all citizens. Other laws, including the Dune Protection Act, Coastal Erosion Act, Coastal Erosion Planning and Response Act, and the Coastal Management Program, are intended to protect our coast and insure wise use and development. For additional information on legislation aimed at protecting our coasts, see http://www.texascoast geology.com/papers/coastlawgeol.pdf and the Texas General Land Office Web site (http://www.glo.state.tx.us/coastal.html).

The landward boundary of the public beach easement is the line of vegetation, and that boundary is constantly moving landward as a result of rising relative sea level and a shortage of sand within the coastal system. As the shoreline and vegetation line advance landward and overtake private property, the Attorney General's Office is responsible for enforcing the law and insuring that structures limiting beach access are removed. This is a tough law, but without it beach access will eventually vanish. The Texas General Land Office (GLO) is the agency responsible for monitoring the coast to see that beach access does not diminish over time. The Coastal Coordination Council, made up of citizens, engineers, and scientists, advises the GLO

Signs like this one are posted along the coast, but we should not take beach access for granted. Not every coastal state provides good public access to its beaches. We also need to guard against having this right taken away by allowing our beaches to shrink.

on management of the beach and dune system. Monitoring changes in the position of the Gulf and bay shorelines is the responsibility of the Bureau of Economic Geology (http://www.beg.uTexas.edu/coastal/intro.htm).

The Coastal Erosion Planning and Response Act (CEPRA) was implemented to reduce the effects of and to understand the processes of coastal erosion and its impact on public beaches, natural resources, coastal development, and public infrastructure (http://www.glo.state.tx.us/coastal). This act further authorizes the GLO to implement a comprehensive coastal erosion program that can include design, funding, building, and maintaining projects to help combat coastal erosion. To date the funds have been used mainly for dune restoration and shoreline stabilization.

The National Oceanic and Atmospheric Administration (NOAA) National Marine Fisheries Service Galveston Laboratory (http://galveston.ssp.nmfs.gov/) is the agency with primary responsibility for wetlands protection. The agency's charge in this matter comes from the Clean Water Act Protection of Wetlands: Executive Order No. 11990, which directs federal agencies to avoid long- and short-term adverse impacts associated with the destruction or modification of wetlands. This law also requires that federal agencies avoid direct or indirect support of new construction in wetlands wherever there is a practicable alternative.

In 1988 the Galveston Bay complex, which includes an area much larger than Galveston Bay proper, was included within the National Estuary Program. The Galveston Bay National Estuary Program has as its main objective the prevention of habitat loss and alteration (http://www.epa.gov/owow/estuaries/programs/gb.htm). The U.S. Army Corps of Engineers is responsible for issuing permits for certain coastal projects, such as those that would infringe on wetlands. The U.S. Geological Survey is also currently engaged in coastal projects in Texas. I have surely omitted some agencies. With so many responsible for safeguarding our coast, communication becomes a serious issue.

Fortunately there are certain watchdog groups that protect our right to beach access and that work hard to protect coastal wildlife and their habitats. The Texas chapter of the Surfrider Foundation (http://www.surfrider.org/) and the Texas Open Beaches Advocates (http://texasopenbeaches.org/) have been the most assertive advocates of beach access. The Galveston Bay Foundation has been instrumental in preserving our wetlands and bays Their mission is to "preserve, protect and enhance the natural resources of the Galveston Bays estuarine system and its tributaries for present users and for posterity." Audubon Texas (http://www.tx.audubon .org.) keeps a watchful and caring eye on our feathered neighbors, and the Sierra Club Galveston Group fights to preserve the natural state of the island and prevent overdevelopment. These groups are fighting a big battle and deserve our support.

How much development can the coast withstand?

As someone who spends every opportunity on the coast, I marvel at how undeveloped the upper Texas coast is, especially considering its proximity to Houston. That is changing fast. New developments are popping up all along Galveston Island, Bolivar Peninsula, and Follets Island. People have finally caught on—the upper Texas coast is a great place to spend time. I hope Galveston Island is never like my old home, Gulf Shores, Alabama. There you can drive for miles and never see the Gulf, and if you try to walk on the beach you can do so only at public parks, which are few and far between. Gulf Shores is not unique; right to beach access is in many states limited to state and federal properties. I hope Texas' open beach access never changes. I also hope there are always cows on Galveston Island and that the sandhill cranes continue to return for the winter. The reality is that we can't stop development on the coast, but we must develop wisely. We should discourage development in areas that are prone to storm breaching and over-wash, development that does not allow wetlands to migrate landward, or construction of houses on freshwater marshes.

The problem is that development occurs project by project, with each project facing its own permit process. Most developers know how to win permit disputes. They simply line up their expert witnesses. The sad part is that they can easily find "experts" to say that their actions will have no impact on the wetlands or that their new development is not prone to storm breaching, even though most serious scholars would disagree. In my profession we call these so-called experts hired guns. I have not met many who I thought were terribly competent, but they often go unchallenged, which is how they get to be considered experts. In the future,

On the far west end of Galveston Island is the largest, most costly single development project ever undertaken on the upper Texas coast. It sits in an area that is highly prone to storm over-wash and infringes on wetlands.

government agencies need to make sure they bring in their own qualified and unbiased experts when considering development projects that may negatively impact the coast. But how can government officials know which projects may have negative impacts? They should adopt codes that prevent unwise development. These codes should be based on geohazard maps of the coast (http://www.bcg.utexas.edu/coastal/coastal01.htm). Such maps designate areas where development will have either adverse impacts on the environment or pose a safety threat to those who inhabit the development. This seems like a sensible thing to do, but I have learned through firsthand experience that adopting such a set of standards is not

Aerial photograph showing old and new developments on the far west end of the island. Note the linear channels on the bay side of the photograph where storms have breached the barrier. This area is currently under development, with homesites located right at the edge of the current wetlands, leaving no room for them to grow in the future. In the lower right corner of the photograph are roads that were constructed during the 1980s for a new development, San Luis Shores. The entrance road has since been overtaken by the vegetation line, so that had homes been built here they would now be on the beach. Photo from GlobeXplorer.

easy. Still, progress is being made, at least for Galveston Island, and the Bureau of Economic Geology has recently completed a geohazard map for the island. The state of Texas intends to follow up on this by constructing geohazard maps for the entire Texas coast. Only after such maps are available will we succeed in stopping irresponsible development.

What can and should we do to protect our coast from overdevelopment?

Conclusions

From a geological standpoint, the upper Texas coast is subject to constant change because it is subsiding relatively fast and because sand sup-

Give those wetlands a little room to grow. This development plan for the bay side of the far west end of Galveston Island leaves little room for wetlands to migrate.

ply to the coast is minimal. Humans have contributed to the instability of the coast, and major storms are gradually nipping away at the shoreline and bay shore. So we must use common sense in developing the coast. The destruction of the Brazos Delta stands as an example of how rapidly the coastal system responds to human tampering.

As far as beaches are concerned, they will always retreat landward, and people who build too close to the vegetation line will always pay the ultimate price of seeing their homes destroyed. It is also a sure bet that some homeowners will try to figure out ways to have taxpayers pay for their mistakes. But the only effective way to combat beach erosion is beach nourishment, which will be expensive along most of the upper Texas coast because large sand resources are located too far offshore to be pumped directly onshore. Whether the federal or state governments will ever pay to exploit offshore sand resources is uncertain, but chances are that this will remain a low priority for federal funding for the foreseeable

future. So the limited near-shore sand resources that do exist will have to be used wisely and should probably be reserved for the day they are needed to protect the seawall.

We need to develop our coast wisely because we may not be able to fix what we break. Protection of wetlands, including small freshwater bodies, should remain our highest priority. We must stop nibbling away at the wetlands, and we must give them space to retreat from the rising sea. Without them, the entire ecosystem will be seriously impacted. Adopting a reasonable setback law that prevents construction of houses so close to the beach that they will threaten beach access in a matter of years is just good common sense. Beach access does not mean having to scale obstacles like sand socks or crawl through culverts to get to the beach.

Responsible development of the coast means allowing plenty of space between the shoreline and houses. Responsible government means making sure these things are done.

This aerial view of the west end of Galveston Island shows an example of houses on the Gulf side of the highway that were constructed with a reasonable setback distance from the vegetation line. Also, the new development on the bay side is set back from the bay so that the wetlands have room to migrate.

Walkways over sand socks require constant maintenance. During storms they tend to be destroyed, so that beach access becomes difficult, if not impossible, especially for the aged and handicapped. The sign reads "Keep Off Geo-Tube—Use Walkway at Your Own Risk." Hardly an invitation to enjoy the beach.

Lastly, we need to teach our children about the coast and its vulnerability. Their generation is the one that will have to deal with our mistakes and with the impact of rising sea level. They will be forced to make some very tough decisions because we are not dealing effectively with the problems now.

References

Abdulah, K. C., J. B. Anderson, J. B. Snow, and L. Holdford-Jack. 2004. "The Late Quaternary Brazos and Colorado Deltas, offshore Texas—their evolution and the factors that controlled their deposition." In *Late Quaternary Stratigraphic Evolution of the Northern Gulf of Mexico Basin*, edited by J. B. Anderson and R. H. Fillon, 237–70. Society of Sedimentary Research, Special Publication No. 79.

Anderson, J. B., K. Abdulah, S. Sarzalejo, F. P. Siringan, and M. A. Thomas. 1996. *Late Quaternary Sedimentation and High-Resolution Sequence Stratigraphy of the East Texas, Shelf.* Geological Society Special Publication No. 117, p. 95–124.

Anderson, J. B., A. Rodriguez, K. Abdulah, L. A. Banfield, P. Bart, R. Fillon, H. McKeown, and J. Wellner. 2004. "Late Quaternary stratigraphic evolution of the northern Gulf of Mexico: A synthesis." In *Late Quaternary Stratigraphic Evolution of the Northern Gulf of Mexico Basin*, edited by J. B. Anderson and R. H. Fillon, 1–24. Society of Sedimentary Research, Special Publication No. 79.

Anderson, J. B., F. P. Siringan, W. C. Smyth, and M. A. Thomas. 1991. "Episodic nature of Holocene sea level rise and the evolution of Galveston Bay." In *Coastal Depositional Systems in the Gulf of Mexico*, 8–14. Gulf Coast Section Society of Sedimentary Research, Twelfth Annual Research Conference.

Anderson, J. B., F. P. Siringan, M. Taviani, and J. Lawrence. 1991. "Origin and evolution of Sabine Lake, Texas–Louisiana." Transactions of the Gulf Coast Association of Geological Societies 41:12–16.

Anderson, J. B., M. A. Thomas, F. P. Siringan, and W. C. Smyth. 1992. "Quaternary evolution of the East Texas Coast and Continental Shelf." In *Quaternary Coasts of the United States: Marine and Lacustrine Systems, Project No. 274, Quaternary Coastal Evolution*, edited by Charles H. Fletcher III and John F. Wehmiller, 253–63. Special Publication, Vol. 48, Society of Economic Paleontologists and Mineralogists, Tulsa, Okla. .

Bard, E., B. Hamelin, M. Arnold, L. Montaggioni, G. Cabioch, G. Faure, and F. Rougerie. 1996. "Sea level record from Tahiti corals and the timing of deglacial meltwater discharge." *Nature* 382:241–44.

Berger, E. 2005. "Rising growth, sinking fortunes: Island is losing land to sea levels and subsidence at higher rate than had been thought." *Houston Chronicle.* Dec. 24.

Bernard, H. A., C. F. Major Jr., B. S. Parrot, and R. J. LeBlanc. 1970. "Recent sediments of southeast Texas: A field guide to the Brazos alluvial and deltaic plains and the Galveston barrier island complex." The University of Texas at Austin Bureau of Economic Geology Guidebook no. 11.

Brown, L. F., Jr., R. A. Morton, J. H. McGowen, C. W. Kreitler, and W. L. Fisher. 1974. "Natural hazards of the Texas coastal zone." The University of Texas Bureau of Economic Geology Special Report.

Brunn, P. 1962. "Sea level as a cause of shore erosion." *Journal of Waterways and Harbors Division, ASCE*, 88:117–30.

Cazenave, A. 2006. "Present-day sea level rise and climate change: observations and causes." In *Sea Level Changes: Records, Processes, and Modeling*, edited by G. Camion, A. Droxler, C. Fulthrope, and K. Miller, 27. SEALAIX 06, Giens, France.

Davis, R. A. 1997. *The Evolution of Coast.* New York: Scientific American Library.

Davis, R. A., and D. M. Fitzgerald. 2004. *Beaches and Coasts.* Malden, Mass.: Blackwell Publishing.

Douglas, B. C., M. S. Kearney, and S. P. Leatherman, eds. 2001. *Sea Level Rise: History and Consequences.* International Geophysics Series, vol. 75. New York: Academic Press.

Dupre, W. R. 1985. "Geologic effects of Hurricane Alicia (August 18, 1983) on the upper Texas Coast." Transactions of the Gulf Coast Geological Societies 35:353–59.

Finkl, Charles W., Jr., Jeffrey L. Andrews, and Lindino Benedet, Coastal Planning and Engineering Inc. 2004. *Reconnaissance Geotechnical and Geophysical Investigations to Identify Offshore Sand Sources for Beach Nourishment in Galveston and Jefferson Counties, TX.* Prepared for Galveston County, Jefferson County, U.S. Army Corps of Engineers, Galveston District. July. (CPE address: 2481 N.W., Boca Raton Blvd., Boca Raton, FL 33431).

Gabrysch, R. K., and C. W. Bonnet. 1975. *Land-surface Subsidence Resulting from Ground-water Withdrawals in the Houston-Galveston Region, Texas,* Texas Water Development Board Report 188.

Gibeaut, J. C., W. A. White, and T. A. Tremblay. 2000. "Coastal hazards atlas of Texas: A tool for hurricane preparedness and coastal management." Vol. 1, *The Southeast Coast: A Report to the Texas Coastal Coordination Council.* Bureau of Economic Geology, Austin, Tex.

Gornitz, V., and S. Lebedeff. 1987. "Global sea-level change during the past century." In SEPM Special Publication 41:10, fig. 7.

Houghton, J. 1994 *Global Warming: The Complete Briefing.* 2nd ed. New York: Cambridge University Press.

Hoyt, J. 1967. "Barrier islands formation." *Bulletin of the Geological Society of America* 82:59–66.

Intergovernmental Panel on Climate Change (IPCC). 2001. *Climate Change 2001.* Synthesis Report. Ed. R. T. Watson and Core Writing Team. Cambridge, U.K.: Cambridge University Press.

Louisiana Department of Natural Resources. 2003. *Holly Beach Sand Management Summary Data and Graphics, CS-31.* Coastal Restoration Division, Baton Rouge, La.

Morton, R. A. 1997. *Gulf Shoreline Movement between Sabine Pass and the Brazos River, Texas: 1974–1996,* The University of Texas at Austin, Bureau of Economic Geology Geological Circular 97–3.

Morton, R. A., and J. G. Gibeaut. 1995. *Physical and Environmental Assessment of Sand Resources, Sabine and Heald Banks.* Final Report to U.S. Department of Interior Minerals Management Services, Bureau of Economic Geology, Austin, Tex.

Morton, R. A., J. C. Gibeaut, and J. G. Paine. 1995. "Meso-scale transfer of sand during and after storms: Implications for prediction of shoreline movement." *Marine Geology* 126:161–79.

Morton, R. A., and J. G. Paine. 1985. *Beach and vegetation line changes at Galveston Island, Texas Erosion, Deposition and Recovery from Hurricane Alicia.* The University of Texas at Austin, Bureau of Economic Geology Geological Circular 85–5.

Morton, R. A., J. G. Paine, and J. C. Gibeaut. 1994. "Stages and duration of post-storm recovery, southeastern Texas Coast, U.S.A." *Journal of Coastal Research* 10:884–908.

Morton, R. A., O. H. Pilkey Jr., O. H. Pilkey Sr., and W. J. Neal. 1983. *Living with the Texas Shore* Durham, N.C.: Duke University Press.

Moya, Juan. 2006. Personal communication. Jan. 20.

Paine, J. G., and R. A. Morton. 1986. *Historical Shoreline Changes in Trinity, Galveston, West and East Bays, Texas Gulf Coast.* The University of Texas at Austin, Bureau of Economic Geology Geological Circular 86–3.

Rodriguez, A., J. B. Anderson, and J. Bradford. 1998. "Holocene tidal deltas of the Trinity Valley: Analogs for exploration and production." Transactions of the Gulf Coast Association of Geological Societies, 48:373–80.

Rodriguez, A. B., J. B. Anderson, and M. Hamilton. 2000. "Evolution and facies architecture of the modern Brazos Delta, Texas: Wave versus flood influence." *Journal Sedimentary Research* 70:283–95.

Rodriguez, A. B., J. B. Anderson, and A. R. Simms. 2005. "Terrace inundation as an autocyclic mechanism for parasequence formation: Galveston Estuary, Texas, U.S.A." *Journal of Sedimentary Research* 75:608–620.

Rodriguez, A. B., J. B. Anderson, F. P. Siringan, and M. Taviani. 2000. "Sedimentary facies and genesis of Holocene sand banks on the East Texas inner continental shelf." In *Isolated Shallow Marine Sand Bodies,* edited by J. Sneddin and K. Bergman, 165–78. SEPM Special Publication No. 64.

Rodriguez, A. B., J. B. Anderson, F. P. Siringan, and M. Taviani. 2004. "Holocene evolution of the east Texas coast and inner continental shelf: Along-strike variability in coastal retreat rates." *Journal of Sedimentary Research* 74:406–422.

Rodriguez, A. B., M. Fassell, and J. B. Anderson. 2001. "Variations in shoreface progradation and ravinement along the Texas coast, Gulf of Mexico." *Sedimentology* 48:837–53.

Shinkle, K. D., and R. K. Dokka. 2004. *Rates of Vertical Displacement at Benchmarks in the Lower Mississippi Valley and the Northern Gulf Coast.* NOAA Technical Report NOS/NGS 50, NOAA, Silver Spring, Md.

Siringan, F. P., and J. B. Anderson. 1993. "Seismic facies, architecture, and evolution of the Bolivar Roads tidal inlet/delta complex, East Texas Gulf Coast." *Journal of Sedimentary Petrology* B64:99–110.

Siringan, F. P., and J. B. Anderson. 1994. "Modern shoreface and inner-shelf storm deposits off the East Texas Coast, Gulf of Mexico." *Journal of Sedimentary Research* B64:798–808.

Swift, D. J. P. 1975. "Tidal sand ridges and shoal-retreat massifs." *Marine Geology* 18:105–134.

Texas General Land Office. *Dune Protection and Improvement Manual for the Texas Gulf Coast.* 3rd ed. *Coastal Issues.* Austin, Texas.

Thomas, M. A., and J. B. Anderson. 1994. "Sea-Level Controls on the Facies Architecture of the Trinity/Sabine Incised-Valley System, Texas Continental Shelf." In *Incised Valley Systems: Origin and Sedimentary Sequences,* edited by R. Dalrymple, R. Boyd, and B. A. Zaitlin, 63–82. SEPM Special Publication 51, Tulsa, Okla.

White, W. A., and R. A. Morton. 1997. "Wetland losses related to fault movement and hydrocarbon production, southeastern Texas Coast." *Journal of Coastal Research* 13:1305–20.

White, W. A., R. A. Morton, and C. W. Holmes. 2002. "A comparison of factors controlling sedimentation rates and wetland loss in fluvial-deltaic systems, Texas Gulf Coast." *Geomorphology* 44:47–66.

White, W. A., and T. A. Tremblay. 1995. "Submergence of wetlands as a result of human-induced subsidence and faulting along the upper Texas Gulf Coast." *Journal of Coastal Research* 11:788–807.

White, W. A., T. A. Tremblay, E. G. Wermund Jr., and L. R. Handley. 1993. *Trends and Status of Wetland and Aquatic Habitats in the Galveston Bay System, Texas.* The Galveston Bay National Estuary Program, Publication GBNEP-31.

Williams, S. J., D. A. Prins, and E. P. Meisburger. 1979. *Sediment Distribution, Sand Resources, and Geologic Character off Galveston Island, Texas.* U.S. Army Corps of Engineers, Report No. 79–4.

Index